Forensic DNA Analysis

A Laboratory Manual

J. Thomas McClintock

 CRC Press
Taylor & Francis Group
Boca Raton   London   New York

CRC Press is an imprint of the
Taylor & Francis Group, an **informa** business

CRC Press
Taylor & Francis Group
6000 Broken Sound Parkway NW, Suite 300
Boca Raton, FL 33487-2742

© 2008 by Taylor & Francis Group, LLC
CRC Press is an imprint of Taylor & Francis Group, an Informa business

Library of Congress Cataloging-in-Publication Data

McClintock, J. Thomas.
    Forensic DNA analysis : a laboratory manual / J. Thomas McClintock.
       p. ; cm.
    Includes bibliographical references and index.
    ISBN 978-1-4200-6329-5 (hardcover : alk. paper)
    1. DNA--Analysis--Laboratory manuals. 2. DNA fingerprinting--Laboratory manuals. 3. Forensic genetics--Laboratory manuals. I. Title.
    [DNLM: 1. DNA Fingerprinting--Laboratory Manuals. 2. Forensic Medicine--methods--Laboratory Manuals. W 625 M478f 2008]

RA1057.55.M33 2008
614'.1078--dc22   2007045327

**Visit the Taylor & Francis Web site at
http://www.taylorandfrancis.com**

**and the CRC Press Web site at
http://www.crcpress.com**

Contents

List of Figures

List of Tables

Welcome to the Forensic DNA Analysis Laboratory

DNA typing has revolutionized criminal investigations and has become a powerful tool in the identification of individuals in criminal and paternity cases. In the past few years, the general public has become familiar with forensic DNA typing based on exposure from media coverage (e.g., the O. J. Simpson trial, the President Bill Clinton and Monica Lewinsky scandal, and the identification of individuals killed in the September 11, 2001 attacks on the World Trade Center in New York City and the Pentagon in Arlington, Virginia) and television programs (e.g., *Forensic Files* and *CSI: Miami*). Although these cases have generated widespread media attention, they represent only a small fraction of the thousands of forensic DNA and paternity cases that are conducted by public and private laboratories in the United States and abroad.

The purpose of this *Forensic DNA Analysis Laboratory Manual* is to introduce the student to the science of DNA typing methods by focusing on basic techniques used in forensic DNA laboratories. This laboratory manual is designed to provide the student with a fundamental understanding of forensic DNA analysis as well as a thorough background of the molecular techniques used to determine an individual's identity or parental lineage. This manual is intended to challenge the student with the methodology of the investigation in DNA typing, help the student develop an understanding of the scientific principles involved in DNA analysis, and ensure the student is able to analyze and interpret the data that are generated in each exercise with clarity and confidence.

The exercises in this laboratory manual have been organized to first provide an overview of forensic DNA analysis, the sources or types of biological material used in DNA analysis, and then the background principles and practical methodology for a specific DNA typing technique. In some exercises, the protocols have been adapted from methods and protocols used in federal, state, and private forensic laboratories. Each exercise is designed to simulate human forensic testing but can also be used to simulate a wide range of applications for genetic analysis. The actual scenario employed in each exercise is up to the discretion of the course instructor. Lastly, an extensive glossary has been included to assist students with DNA typing terminology as well as basic terms used in molecular biology.

Compiled below is a brief history of forensic DNA typing. Since DNA testing was first introduced in the United States in 1986, it has been used in thousands of cases. However, the list below highlights specific events or developments in forensic DNA analysis as well as those cases brought to the attention of the general public by media exposure.

Brief History of Forensic DNA Typing

1980   Ray White describes first polymorphic RFLP marker.
1985   Alec Jeffreys discovers multilocus VNTR probes.
1985   First paper on Polymerase Chain Reaction (PCR)

1986   DNA testing goes public (Cellmark and Lifecodes).

1986   First RFLP case in the U.S. (Florida vs. Tommy Lee Andrews)

1988   FBI starts DNA casework (RFLP).

1989   The Technical Working Group on DNA Analysis Methods (TWGDAM) established

1991   First Short Tandem Repeats (STR) paper

1992   NRC I Report "DNA Technology in Forensic Science"

1993   First STR kit available

1995   Forensic Science Service (FSS) starts UK DNA database

1995   O.J. Simpson trial; public becomes aware of DNA.

1996   NRC II Report "The Evaluation of Forensic DNA Evidence"

1996   First use of mitochondrial DNA test in a U.S. criminal trial (Tennessee v. Ware).

1998   FBI launches Combined DNA Index System (CODIS) database.

1998   Establishment of Quality Assurance standards for forensic DNA testing laboratories through the DNA Advisory Board.

1998   Kenneth Starr investigates allegations of President Clinton's sexual relationship with White House intern Monica Lewinsky.

1999   Multiplex STRs are validated.

1999   The decision in State v. Ware (1996) was upheld by an appellate court.

2002   Division of Forensic Science Laboratory in the Commonwealth of Virginia became the first state laboratory to mark 1,000 "cold hits" from its DNA database.

2003   A field DNA test was completed to provide preliminary confirmation of the identification of Saddam Hussein less than 24 hrs after his capture. A full test performed in the laboratory, provided confirmation.

2003   The National Institute of Standards and Technology develops a "mini-STR assay" to allow the remains from 16 additional victims from the September 11, 2001 attacks on the WTC to be positively identified.

2006   Members of the Duke University men's lacrosse team arrested and accused of raping a female exotic dancer. Samples collected from the dancer and the men's lacrosse team for DNA analysis.

2006   DNA testing failed to connect any members of the Duke University men's lacrosse team to the alleged sexual assault of an exotic dancer.

2007   Prosecutor handling the Duke case is forced to recuse himself. North Carolina's attorney general declared three former Duke University lacrosse players who had been accused of gang-raping a stripper innocent of all charges, ending a prosecution that provoked bitter debate over race, class, and the tactics of the Durham County district attorney.

2007   Applied Biosystems introduces the first DNA testing kit for analyzing degraded or limited DNA

2007   FSS in the UK uses laser microdissection (LMD), which enables single cells to be extracted from a microscope slide, with fluorescence in situ hybridization (FISH), a method to highlight chromosomes, to distinguish between male (XY chromosomes) female (XX chromosomes) cells.

It is hoped that this manual will develop the curiosity and confidence of the student to further explore questions and issues involving forensic science investigations. I look forward to teaching you the techniques and applications in forensic DNA analysis.

Laboratory Rules

1. No eating, drinking, smoking, applying cosmetics, or handling contacts in the laboratory at **ANY** time.

2. No pipetting by mouth. Use a pipettor at all times.

3. Minimize splashing and production of aerosols.

4. Store all books, backpacks, cell phones and other electronic devices, purses, coats, and so on in the cabinet of your laboratory bench (or designated area). Only your laboratory notebook should be on the bench.

5. Do not place pencils, pens, or any other object into your mouth while in the laboratory.

6. **NEVER** take any reagents, samples, or cell cultures out of the laboratory.

7. Notify the laboratory instructor immediately of any spills, of any accidents, or if you cut or injure yourself.

8. In most instances, you will be wearing disposable gloves. Be extra careful when handling reagents or chemicals to eliminate skin contact.

9. Wash your hands at the beginning and at the end of the laboratory exercises.

10. Clean your laboratory bench with dilute alcohol before you begin work and when you have completed the laboratory exercise.

11. Laboratory coats are not required. However, in forensic laboratories, laboratory coats and disposable gloves must be worn.

12. Children are not allowed in the laboratory.

13. Familiarize yourself with the location of the eye wash station, the fire extinguisher, and the fire blanket.

Chapter 1

An Overview of Forensic DNA Analysis

Introduction

Everyone has a unique set of fingerprints. As with a person's fingerprint, no two individuals share the same genetic makeup. This genetic makeup, which is the hereditary blueprint imparted to us by our parents, is stored in the chemical deoxyribonucleic acid (DNA), the basic molecule of life. Examination of DNA from individuals, other than identical twins, has shown that variations exist and that a specific DNA pattern or profile can be associated with an individual. These DNA profiles have revolutionized criminal investigations and have become powerful tools in the identification of individuals in criminal and paternity cases.

Restriction Fragment Length Polymorphism

The first widespread use of DNA tests involved restriction fragment length polymorphism (RFLP) analysis, a test designed to detect variations in human DNA. In the RFLP method, DNA is isolated from a biological specimen (e.g., blood, semen, or vaginal swabs) and cut by an enzyme into pieces called restriction fragments. The DNA fragments are separated by size into discrete bands by gel electrophoresis, transferred onto a membrane, and identified using probes (known DNA sequences that are "tagged" with a chemical tracer). The resulting DNA profile is visualized by exposing the membrane to a piece of X-ray film allowing the scientist to determine which specific fragments the probe identified among the thousands in a sample of human DNA. A "match" is made when similar DNA profiles are observed between an evidentiary sample and those from a known sample (e.g., DNA from a victim or suspect). A determination is then made as to the probability that a person selected at random from a given population would match the evidence sample as well as the suspect. The entire analysis may require several weeks for completion.

Polymerase Chain Reaction–Based Tests

In instances when the evidentiary sample contains an insufficient quantity of DNA or the DNA is degraded, a polymerase chain reaction (PCR)–based test is used to obtain a DNA profile. The PCR-based tests generally provide rapid results that can serve as an alternative or as a complement to other DNA tests. The first step in the PCR process involves the isolation of DNA from a biological specimen (e.g., blood, semen, saliva, or fingernail clippings). Next, the PCR amplification technique is used to produce millions of copies of a specific portion of a targeted DNA segment. The PCR amplification procedure is comparable to a photocopying machine, only it is at the molecular level. The amplified PCR products are then identified by either the addition of known DNA probes (e.g., DQA1 and PM test kits) or separation by gel electrophoresis (D1S80, short tandem repeat [STR], and amelogenin [gender] analyses) followed by chemical staining. Such detection procedures eliminate the need for critically sensitive DNA probes, thus reducing the analysis time from several weeks to 24–48 hours. The resulting DNA profiles are routinely interpreted by direct comparison to DNA standards. Probability calculations are determined based upon classical population genetic principles.

Mitochondrial DNA Analysis

Mitochondrial DNA (mtDNA) typing is increasingly used in human identity testing when biological evidence may be degraded, when quantities of the samples in question are limited, or when nuclear DNA typing is not an option. Biological sources of mtDNA include hairs, bones, and teeth. In humans, mtDNA is inherited strictly from the mother. Consequently, mtDNA analysis cannot discriminate between maternally related individuals (e.g., mother and daughter, or brother and sister). However, this unique characteristic of mtDNA is beneficial for missing person cases when mtDNA samples can be compared to samples provided by the maternal relative of the missing person.

In humans, the mtDNA genome is approximately 16,000 bases (A, T, G, and C) in length containing a "control region" with two highly polymorphic regions. These two regions, termed Hypervariable Region 1 (HV1) and Hypervariable Region 2 (HV2), are 342 and 268 base pairs (bp) in length, respectively, and are highly variable within the human population. This sequence (the specific order of bases along a DNA strand) variability in either region provides an attractive target for forensic identification studies. Moreover, because human cells contain several hundred copies of mtDNA, substantially more template DNA is available for amplification than nuclear DNA.

Mitochondrial DNA typing begins with the extraction of mtDNA followed by PCR amplification of the hypervariable regions. The amplified mtDNA is purified and subjected to sequencing (Sanger et al., 1977), with the final products containing a fluorescently labeled base at the end position. The products from the sequencing reaction are separated, based on their length, by gel or capillary electrophoresis. The resulting sequences or profiles are then compared to sequences of a known reference sample to determine differences and similarities between samples (Anderson et al., 1981; Andrews et al., 1999). Samples are not excluded as originating from the same source if each base (A, T, G, or C) at every position along the hypervariable regions is similar. However, due to the size of the mtDNA database and to the unknown number of mtDNA sequences in the human population, a reliable frequency estimate is not provided. Consequently, mtDNA sequencing is becoming known as an exclusionary tool as well as a technique to complement other human identification procedures.

Chapter 2

Types of Biological Samples

With the exception of white blood cells, DNA is found in every human cell. Consequently, DNA is present in a variety of body fluids and tissues that have been demonstrated to be suitable for DNA typing. However, if the sample or evidence collection is performed improperly the sample's integrity may be compromised, leading to contamination and/or degradation. Improper handling procedures during storage and transport from the crime scene to the laboratory can result in samples unfit for analysis. The importance of sample integrity cannot be overemphasized because ambiguous data or information can compromise the investigation and/or the outcome of the case.

In the past, DNA typing tests such as restriction fragment length polymorphism (RFLP) were successful in generating complete DNA profiles provided that adequate and nondegraded samples were utilized. The introduction of the polymerase chain reaction (PCR) in the mid- to late 1980s extended the range of possible samples available for DNA analysis regardless of their condition. Some of the biological samples that have been tested successfully with PCR-based typing methods are listed in Table 1. The minimum or corresponding amount of DNA available from each biological sample is also shown.

Prior to DNA isolation and typing, biological samples must first be collected either from a known contributor (victim and/or suspect) or from the crime scene (evidentiary sample). Once the sample is collected, the DNA is extracted and subjected to PCR analysis. In general, the PCR procedure typically requires as little as 1 nanogram (1 billionth of a gram) of high molecular weight genomic DNA. DNA thought to be degraded can also be subjected to PCR analysis because intact, high molecular weight DNA is not necessary to generate a complete DNA profile. Table 2 illustrates the types of physical evidence collected at crime scenes, the various locations of the DNA, and the biological source.

Biological evidence will attain its full forensic value only when the DNA types can be compared to known profiles obtained from the victims and suspects. The following is a brief description of the various biological samples suitable for DNA typing with accompanying guidance for collection and storage.

Whole Blood

Whole blood from a known source should be collected in a sterile tube containing the preservative EDTA (ethylenediamine tetraacetic acid). In addition to acting as a preservative, EDTA also inhibits the activity of enzymes that are responsible for degrading DNA. Tubes containing blood samples should be stored at

TABLE 1
DNA Contents of Biological Samples

Biological Sample or Source	Approximate DNA Content
Liquid blood	20–40 µg/ml (1 µl from 4–11 × 10³ white blood cells)
Blood stain (1 cm²)	250–500 ng
Semen	150,000–300,000 ng/ml
Postcoital vaginal swab	0–3000 ng/ml
Saliva	1000–10,000 ng/ml
Oral swab	1000–1500 ng
Hair roots	1–750 ng/plucked hair root
Hair (shed)	1–12 ng/hair
Urine	1–20 ng/ml
Bone	3–10 ng/mg bone
Tissue (15 mg)	3–15 µg/mg
Fibroblast cell line	6.5 µg/1 × 10⁶ cells

refrigerated (for short periods of time) or frozen temperatures (for long-term storage). Whole blood can also be "spotted" onto an FTA© collection card (an absorbent cellulose-based paper that contains chemical substances to inhibit bacterial growth and to protect the DNA from enzymatic degradation), allowed to dry, and stored at room temperature for several years.

Bloodstains or Mixed Stains

Garments or clothing containing stains are packaged in a suitable manner (e.g., in a bag and/or box) and transported to the laboratory and stored until analysis. It should be noted that all stained material should be

TABLE 2
Physical Evidence Collected at Crime Scenes

Physical Evidence	Location of DNA	Biological Source of DNA
Used cigarette	Cigarette butt	Saliva
Toothpick	Tips	Saliva
Stamps and envelopes	Licked area	Saliva
Bottle or can	Mouthpiece	Saliva
Used condom	Inside or outside surface	Semen, or vaginal or rectal cells
Blanket, sheet, or pillow	Surface	Semen, sweat, hair, saliva, or urine
Bite mark	Clothing or person's skin	Saliva
Fingernail	Scrapings	Blood, skin, or sweat
Tape or ligature	Inside or outside surface	Skin or a surface
Bullet	Outside surface	Blood, tissue, or skin
Clothing	Surface area	Blood, sweat, or semen
Hat, mask, or bandanna	Inside surface	Sweat, hair, or dandruff
Knife, bat, or similar object	Outside surface or handle	Blood, skin, tissue, or sweat

dried thoroughly prior to packaging and submission to the laboratory for analysis. Short-term storage should be at room temperature in a humidity-controlled room out of direct sunlight. For long-term storage, samples should be stored in a low-temperature, frost-free freezer. Once the packaged material arrives at the laboratory, the removal of the stain from the item is performed by the forensic DNA analyst. Stains comprising a size of approximately 5 mm^2 (half the size of a dime) or greater, and with a volume of less than 5 μl, have been successfully analyzed by current DNA typing techniques.

Hairs

In general, forensic hair analysis involves either head or pubic hair. The collection of 12–24 full-length hairs from the scalp will provide more than enough material for analysis. Hair samples that contain an intact root will provide enough nuclear DNA for short tandem repeat (STR) typing (see Exercise 8). A hair shaft contains sufficient mitochondrial DNA (mtDNA) for successful mtDNA typing (see Exercise 11). As with stains, short-term storage may be at room temperature in a humidity-controlled room and out of direct sunlight. For long-term storage of hair samples, a low-temperature, frost-free freezer is recommended.

Swabs from Biological Material or Inanimate Objects

DNA has been successfully analyzed from swabs containing biological material (e.g., buccal cells from the inside of the cheek or epithelial cells from a vaginal swab) and from swabs of various inanimate objects such as cigarettes, envelopes, soda cans, and stamps. Using a sterile and moistened cotton swab, the area in question is "swabbed," then the swab is dried and placed in a container for storage or placed in a vial containing a small volume of sterile solution such as 1X TE buffer for short-term storage. Swabs containing the biological material for analysis are stored as recommended as with stains.

Bone, Teeth, and Tissue

In some instances, the probability of obtaining a sufficient quantity or quality of biological material such as blood or semen for DNA typing will be low. This insufficient quality or quantity of material may be due to sample degradation, availability, or even accessibility. In these cases, samples such as bones, teeth, skin, and/or muscle tissue will usually provide sufficient DNA for analysis. Generally, a 1 cm^2 section (slightly smaller than a dime) of such biological material is suitable for testing. Following collection, the samples should be frozen and transported to the laboratory on ice. Upon arrival at the laboratory, the samples should be kept frozen until the DNA typing analysis begins. To ensure sample integrity, avoid multiple freeze-thaw conditions.

Paraffin-Embedded Tissues, Smears, or Slides

When known biological samples are unavailable, but needed, investigation into local medical facilities may yield a source: specimens collected from biopsies or surgical procedures that were processed, analyzed, and stored. Such specimens may include vaginal or pap smears or histological sections. Histological sections contain samples fixed in formalin, processed, and embedded in paraffin. The paraffin "blocks" are sectioned using a microtome and placed on polylysine-coated slides. Usually, a minimum of one block, smear, or slide is typically necessary for a successful DNA typing analysis. Histological sections, smears, or slides may be stored at room temperature indefinitely prior to analysis.

Semen and Sperm

Sperm specimens, collected from a vaginal swab, will contain epithelial (skin) cells from the female and, in some instances, from the male. Sperm can be preferentially separated from the rest of the material in such a mixture using specific extraction methods discussed in Exercise 1 (see the section entitled "Differential Extraction"). Samples or stains that are thought to, or have been shown to, contain only spermatozoa can be collected and processed as described above (see "Bloodstains or Mixed Stains"). Sperm specimens should be stored frozen prior to analysis. To ensure sample integrity, avoid multiple freeze-thaw conditions.

Urine

Similar to perspiration and sebaceous oils, urine when concentrated will contain a sufficient amount of epithelial cells to generate a DNA profile. Ideally, a minimum sample volume of 10 ml is required for analysis, with an optimum approaching 30 ml. However, a sufficient number of cells might be obtained from a stain or swab of a known source. For long-term storage, the specimen should be stored frozen. Specimens can be stored in refrigerated temperatures for short periods of time prior to analysis. To maintain specimen integrity, it is critical to avoid multiple freeze-thaw conditions.

3

Exercise 1
DNA Extraction

Introduction

There are a number of different approaches for the isolation of genomic DNA. Each procedure begins with some form of cellular lysis, followed by deproteinization and recovery of DNA. The main differences between the various approaches lie in the extent of deproteinization and the size of the DNA isolated. In addition, the isolation or extraction of DNA will vary according to the type of biological sample, the amount of evidence or biological sample, and the type of cell(s) present in the sample.

DNA must first be separated from the rest of the cellular components, as well as from any nonbiological material present. The removal of extraneous substances following cell lysis minimizes sample (DNA) degradation due to cellular enzymes while ensuring maximum enzymatic efficiency during the typing procedure.

Objectives

In this exercise, you will learn different isolation techniques used in forensic DNA analysis to extract DNA from various biological sources. The techniques used in this exercise are representative examples of modern techniques used in forensic laboratories to isolate whole genomic DNA from known samples as well as from evidentiary samples. The extraction procedures described below are relatively brief and easy to perform. The optimum isolation procedure is highlighted for each sample or cell type. When completed, the DNA isolated in this exercise can be utilized in the subsequent exercises.

Extraction Methods

The use of disposable gloves and aerosol-resistant pipet tips is highly recommended to prevent cross-contamination. A helpful organizational sheet is provided at the end of the exercise to record data and other necessary information.

1. Chelex Extraction

When a minimal amount of sample is available (e.g., a spot of blood), the Chelex extraction method has been used. The sample is boiled in a solution containing minute beads of a chemical called Chelex. The boiling causes the cells to lyse, releasing the DNA. The Chelex binds to the extraneous cellular material, and the entire "complex" is removed by centrifugation, leaving the DNA in the supernatant. Because the high temperatures disrupt the two strands of the DNA, generating single-stranded molecules, this extraction process is generally reserved for PCR-based typing techniques.

Equipment and Material

1. 15 ml sterile polypropylene test tube
2. Sterile 1.5 ml Eppendorf or microcentrifuge tubes
3. 5 ml pipettor with sterile tips
4. Adjustable-volume digital micropipets (20–1000 µl range)
5. 10% suspension of Chelex resin beads
6. Aerosol-resistant pipet tips
7. Cultured human cell lines (see Exercise 7 for details)
8. 1.5 ml test tube holder
9. 0.9% saline
10. Sterile cotton swabs
11. Disposable gloves
12. Boiling water bath in a 1000 ml beaker
13. Ice in buckets
14. Tabletop clinical centrifuge
15. Microcentrifuge

Procedure

Collection of Cells (e.g., Buccal Cells, Liquid Blood, and Cultured Human Cells)

1. Label a 15 ml polypropylene test tube and the top of a 1.5 ml Eppendorf tube (also referred to as a microcentrifuge tube) with your name and any other appropriate information.
2. Pipet 10 ml of suspended cells (maximum 5×10^6 cells/ml) or liquid sample into the polypropylene test tube (for harvesting cultured human cells, see the "Salting Out" procedure below, steps 1–6). For buccal cells, rinse your mouth with 10 ml of saline solution and vigorously swish against your cheeks for 10 sec. Expel saline solution back into the labeled 15 ml polypropylene test tube over the sink.

Or,

if sterile swabs are available, place the swab inside your mouth and press it firmly against the inside of your cheek. Roll the swab back and forth over the inside surface of your cheek at least 10 times. Repeat on the other cheek. Place the swab into a labeled 15 ml test tube containing 10 ml saline solution.

Concentrate Cells by Centrifugation

3. Centrifuge the samples at 300 × g for 5 min. The cells form a firm pellet below the saline supernatant. **SAVE THE PELLET AND DISCARD THE SUPERNATANT** by decanting into the sink with running water, taking care not to disturb the cell pellet at the bottom of the tube.

4. Add 500 μl of Chelex beads into the 15 ml test tube containing the cell pellets. Resuspend the cell pellet either by slowly pipetting "in and out" several times or by tapping with your finger.

5. Transfer a 500 μl aliquot of the cell–Chelex slurry into a sterile 1.5 ml Eppendorf tube. Make sure the Eppendorf tube is labeled for identification purposes.

Lysing the Cells and Collecting the DNA

6. Place the capped (closed) Eppendorf tubes in a "float," and place in a boiling water bath for 10 min.

7. After the heat treatment, place the samples on ice for 5 min.

8. Place the Eppendorf tubes containing the lysed cells in a microcentrifuge, and spin at the maximum speed for 1 min. The pellet contains the Chelex beads bound to the denatured proteins. The supernatant contains the DNA.

9. Using a 1000 μl micropipettor with a sterile tip, transfer all of the clear supernatant to a fresh 1.5 ml Eppendorf tube.

10. Label the tube, and place on ice until you are ready to proceed to the next step.

2. Organic Extraction

Organic extraction is a general method used for many situations when stained fabric or clothing is suspected of containing biological material. The stain on the material is cut away from the fabric, soaked in a warm solution (stain extraction buffer) to release the cells from the fabric, and incubated with proteinase K, and the DNA is isolated using organic solvents. The organic extraction method maintains the integrity of the DNA (i.e., large segments are maintained) while "cleaning" the DNA.

Equipment and Material

1. 2.0 ml screw-cap tube

2. 0.5 ml microcentrifuge tube

3. Adjustable-volume digital micropipets (100–1000 μl range)

4. Aerosol-resistant pipet tips

5. Stain extraction buffer

6. Proteinase K (10 mg/ml)

7. Phenol:CHCl$_3$:isoamyl alcohol (25:24:1)

8. Chloroform (CHCl$_3$)

9. Disposable gloves

10. Incubator or water bath at 56°C

11. Vortex mixer

12. Microcentrifuge

Procedure

1. Place the "cutting" from the stained material in a 2.0 ml screw-cap tube.

2. Add 400 µl of stain extraction buffer and 10 µl of proteinase K to each tube. Mix and centrifuge for 2 sec. Incubate the tubes containing the "cutting" at 56°C overnight.

3. Briefly centrifuge the samples for 5 sec.

4. Punch a hole in the bottom of a 0.5 ml microcentrifuge tube. Remove the cutting from the 2.0 ml tube using sterile forceps, and place in the 0.5 ml tube. Place the 0.5 ml tube into the 2.0 ml tube from which the cutting was removed.

5. Centrifuge the 2.0 ml tube containing the 0.5 ml "inserted" tube at maximum speed for 5 min in the microcentrifuge.

6. Remove the 0.5 ml tube, and save the cutting.

7. Replace the screw cap on the 2.0 ml tube.

Note: The following steps should be carried out in an exhaust hood.

8. Add 500 µl to each tube, vortex the tube for 20 sec, and centrifuge in the microcentrifuge for 2 min.

9. Transfer the top aqueous layer containing the DNA to a new 2.0 ml tube. Do not disturb the interface. Dispose of the phenol:CHCl$_3$:isoamyl alcohol solution in the collection tube in a biohazard waste container.

Note: Steps 10 and 11 are optional unless using Centricon concentration.

10. Add 500 µl of CHCl$_3$ to each tube, vortex, and centrifuge for 2 min.

11. Transfer the top aqueous layer to a new 2.0 ml tube. Dispose of the CHCl$_3$ solution in the collection tube in a biohazard waste container. The sample is now ready for precipitation or concentration.

3. "Salting Out"

The salting-out procedure is relatively easy to use with liquid samples (known or evidentiary samples) and with cell cultures that might be used as "mock" evidence samples or as controls. The salting-out DNA isolation procedure involves the preferential hydrolysis and precipitation of cellular proteins. The protein-free genomic DNA is subsequently recovered by either method described in Exercise 2 or 3.

Equipment and Material

1. TE-9 buffer

2. Proteinase K

3. Adjustable-volume digital micropipets (20–200 μl range)

4. Aerosol-resistant pipet tips

5. Cell culture (T-75 flask)

6. 15 ml conical tubes

7. 50 ml conical tubes

8. Disposable gloves

9. 10% SDS

10. Saturated NaCl

11. TE buffer

12. 1X trypsin-EDTA

13. 1X PBS

14. Ice in buckets

15. Inverted microscope

16. Incubator or water bath at 48°C

17. Tabletop clinical centrifuge

Procedure

1. Decant the growth medium from the cell culture flask. Place the growth medium in a 15 ml conical tube, and save for Step 5.

2. Wash the cell monolayer twice with 1X PBS (free of calcium and magnesium), decant, and discard.

3. Add 2 ml of 1X trypsin-EDTA to each cell culture flask. Incubate the flask in the palm of your hands for 30–60 sec. Decant, and discard the trypsin. Under the inverted microscope, observe the "rounding up" of the cells.

Note: In the absence of a microscope, the "rounding up" of the cells can be assessed by holding the flask up to a light source. As the refractive index changes due to the cells rounding up, the bottom of the flask (which the cells are attached to) will appear cloudy or foggy. The extent of this foggy appearance will depend on the degree of cell rounding and the density of the cell population. It is extremely important not to lyse the cells in the presence of trypsin.

4. To completely dislodge the cells, strike the flask against the palm of your hand. It might be necessary to strike the flask several times against your hand to completely dislodge the cells.

5. To inactivate the trypsin, add 5 ml of the saved medium from Step 1.

6. Transfer the cell suspension to a 15 ml conical tube (or suitable centrifuge tube), and centrifuge at 150–200 × g for 3–5 min.

7. Decant the supernatant, and resuspend the cell pellet 4.5 ml in of TE-9 buffer. Add 500 μl of 10% SDS, and invert the tube to mix.

8. Add 125 μl of proteinase K to each tube, and invert to mix. Incubate the samples at least 30 min at 48°C.

9. Add 1.5 ml of saturated NaCl solution to each tube, and shake for 15 sec. The lysate should become and remain cloudy.

10. Centrifuge at 500 × g for 10 min to pellet proteins.

11. Decant the supernatant containing the DNA into a fresh 15 ml tube. Centrifuge for an additional 10 min.

12. Decant the supernatant containing the DNA into a fresh 50 ml conical tube, and place on ice (see Exercise 2 for instructions on how to concentrate the DNA).

4. Differential Extraction

Differential extraction is the method of choice when biological samples are suspected of containing cells from more than one contributor. Differential extraction is commonly used to isolate the male and female components from a sample containing DNA from a male and female contributor. Consequently, differential extraction is used to separate sperm cells from "nonsperm" cells in sexual assault cases. This nonsperm category includes the epithelial cells (or skin cells) found in saliva, buccal swabs, vaginal swabs, urine, and feces. The different properties of sperm cells are exploited to separate them from these "nonsperm" or epithelial cells. The separation of the sperm (sometimes referred to as the male fraction) from the epithelial cells (referred to as the female fraction) provides a DNA profile that is easier for the forensic DNA analyst to interpret in a rape case.

Equipment and Material

1. TNE buffer
2. 20% sarkosyl
3. Sterile deionized water
4. Proteinase K (20 mg/ml)
5. Adjustable-volume digital micropipets (20–1000 µl range)
6. Aerosol-resistant pipet tips
7. 1 M DTT
8. 2.0 ml screw-cap microcentrifuge tube
9. 0.5 ml microcentrifuge tube
10. 2.0 ml screw-cap tubes (conical tubes)
11. Disposable gloves
12. Phenol:CHCl$_3$:isoamyl alcohol (25:24:1)
13. Stain extraction buffer
14. Vortex mixer
15. Microcentrifuge
16. Incubator or water bath at 37°C and 56°C

Procedure

Extraction of DNA from Mixtures (or Mixed Stains)

1. The questioned sample containing a stain thought to contain a mixture of sperm and epithelial cells is placed in a 2.0 ml screw-capped Eppendorf or microcentrifuge tube.

2. Mild detergents are then added to remove the stain containing the cells from the material.

Add:

400 µl TNE

25 µl 20% sarkosyl

75 µl sterile deionized water

5 µl 20 mg/ml proteinase K

3. Mix the sample, and centrifuge for 2 sec. Incubate the sample at 37°C for 2 hrs.

4. Punch a hole in the bottom of a 0.5 ml microcentrifuge tube. Place stained swab into the 0.5 ml tube, and place the 0.5 ml tube into the 2.0 ml tube from which the stained swab was removed. Align the tab on the 0.5 ml tube with the case number label on the 2.0 ml tube.

5. Spin the tube for 5 min.

6. Remove the 0.5 ml tube, and place the swab in a clean 2.0 ml screw-capped tube. This is Fraction 2 (F2).

7. Transfer the supernatant to a clean 2 ml screw-capped tube. This is Fraction 1 (F1). Set aside. The remaining pellet in the tube is the male fraction (M).

8. Add the following components to the pellet labeled (M):

150 µl TNE

150 µl H_2O

50 µl 20% sarkosyl

40 µl 1M DTT

10 µl 20 mg/ml proteinase K

9. Mix, and incubate at 37°C for 2 hrs.

10. Add 500 µl of stain extraction buffer to the swab (F2).

11. Incubate overnight at 56°C.

Note: The following steps should be carried out in an exhaust hood.

12. Add 400 µl phenol:$CHCl_3$:isoamyl alcohol to each male fraction and 500 µl of the same to each female fraction.

13. Vortex, and centrifuge for 2 min.

14. Transfer the top aqueous layer to a new tube. Dispose of the phenol:$CHCl_3$:isoamyl alcohol solution in the collection tube, and dispose of the test tube in the appropriate biohazard waste container.

Note: Steps 15, 16, and 17 are optional unless using Centricon concentration.

15. Add 500 µl $CHCl_3$ to the tube.

16. Vortex, and spin for 2 min.

17. Transfer the supernatant to a new tube *or* remove the bottom layer and discard. Do not disturb the interface. Dispose of the $CHCl_3$ solution in the collection tube. Dispose of the test tube in an appropriate biohazard waste container. The sample is now ready for precipitation or concentration.

5. DNeasy Blood and Tissue Kit

The DNeasy Blood and Tissue Kit (QIAGEN, Inc., Valencia, California) is designed for the rapid isolation, purification, and concentration of total DNA from animal tissue and/or cells. The buffer system, which is supplied by the manufacturer, allows for direct cell lysis followed by selective binding of the DNA to a

FIGURE 1

DNeasy Blood and Tissue Procedure for the Isolation, Purification, and Concentration of DNA. *Source*: Courtesy of QIAGEN, Inc.

silica gel-membrane. The lysate is loaded onto the DNeasy minicolumn and briefly centrifuged. During centrifugation, the DNA binds to the membrane in the minicolumn while contaminants and enzyme inhibitors (e.g., proteins and divalent cations) "pass through" the membrane into a collection tube. Following two wash steps, the DNA is eluted in water or a buffer, and is ready for use (Figure 1). The entire procedure and protocol are presented in Exercise 2.

Results

Before any analysis proceeds, it is important to determine the success of your extraction. It is important to determine the quality and quantity of DNA present and determine if any degradation of the DNA has occurred. The answers to these questions as well as guidelines for interpreting your results are described in Exercise 4.

Samples Extracted

Analyst: _____   Lab Number: _____

Date: _____

Item Number and/or Description	Extraction Method	Purification Method	Item Number and/or Description	Extraction Method	Purification Method

C = Chelex extraction. S = Salting-out extraction. O = Organic extraction. D = Differential extraction.

Comments:

_____

_____

_____

_____

Questions

1. What are the factors that a DNA analyst considers when determining the isolation and extraction procedure to use when analyzing a sample?

2. In your attempt to extract DNA from various samples (both known and evidentiary), several isolation techniques were explored. What technique(s) or method of choice would be used if an evidentiary sample was suspected to contain sperm? Why?

3. In all of the extraction procedures discussed proteinase K and/or a detergent (SDS or sarkosyl) was used in the process. What is the purpose of proteinase K? What is the purpose of the detergents?

Chapter **4**

Exercise 2
Concentration of Extracted DNA

Introduction

Following the removal and purification of the DNA from the sample, the next step is to concentrate the DNA. Various methods exist to concentrate DNA. Two widely used methods includes concentrating extracted DNA by precipitation with ethanol or using a column filtration system (DNeasy Blood and Tissue Kit, QIAGEN, Inc.) to concentrate the DNA. Both techniques are rapid and are quantitative even with nanogram amounts of DNA.

Objective

To concentrate isolated DNA using two different techniques: ethanol precipitation and column filtration (DNeasy Blood and Tissue Kit).

Equipment and Material

1. Phosphate-buffered saline (PBS)
2. 1X TE buffer
3. DNeasy Blood and Tissue Kit
4. 1.5 ml Eppendorf or microcentrifuge tubes
5. Adjustable-volume digital micropipets (2–200 μl range)
6. Aerosol-resistant pipet tips

7. Absolute ethanol (EtOH; 70% and 96–100%)

8. Disposable gloves

9. Ice in buckets, –20°C, or –70°C freezer

10. Incubator or water bath at 37°C, 56°C, and 70°C

11. Microcentrifuge

12. Tabletop clinical centrifuge

Procedure

The use of disposable gloves and aerosol-resistant pipet tips is highly recommended to prevent cross-contamination. Helpful organizational sheets are provided at the end of the exercise.

A. Precipitation of DNA Using Ethanol

1. Estimate the volume of the DNA solution, and add exactly two volumes of ice-cold absolute ethanol (EtOH). The extracted samples from Exercise 1 contain approximately 500 μl of DNA solution. Add 1 ml of cold absolute EtOH to these samples or tubes (containing the aqueous layer). Mix by hand.

2. Place the tube containing the EtOH and sample on ice for 30 min, *or* place the tube in the freezer at –70°C for 30 min. Usually 30–60 min at –20°C is sufficient to allow the DNA precipitate to form.

3. Centrifuge the DNA solution containing the EtOH for 15 min. For most purposes, 10 min using a microcentrifuge at $12,000 \times g$ is sufficient. After centrifugation, decant EtOH.

4. Rinse the DNA pellet with 1 ml of 70% EtOH (room temperature), and centrifuge for 10 min. After centrifugation, decant EtOH.

5. Stand the tube in an inverted position on a layer of absorbent paper until dry (approximately 30 min), or air-dry samples in a secure place.

6. Dissolve the DNA pellet in 36 μl or the desired volume of 1X TE.

7. Resuspend the DNA at 56°C for no more than 2 hrs. To assist in dissolving the pellet, the sample can be heated to 37°C.

8. Store the sample at 4°C in an Eppendorf or microcentrifuge tube.

B. Concentration of DNA Using the DNeasy Blood and Tissue Kit

1. Centrifuge the sample for 5 min at $300 \times g$ at room temperature. If human cell lines (e.g., HepG, HeLa, or K562) are used, centrifuge approximately $1–5 \times 10^6$ cells/ml under the same conditions. Cell lines that are archorage dependent will need to be trypsinized prior to harvesting (see Exercise 1, Section 3, entitled "Salting Out").

2. After centrifugation, decant the supernatant, and add 200 μl of PBS to the pellet.

3. Add 200 μl of "Buffer AL" (provided by the manufacturer of the DNeasy Blood and Tissue Kit) to the resuspended cell pellet (see Figure 1). Mix thoroughly, and incubate at 70°C for 10 min.

4. Add 200 μl of 96–100% EtOH to the sample, and mix thoroughly. The sample containing the Buffer AL should be mixed thoroughly with the EtOH to ensure a homogeneous solution. A white precipitate may form with the addition of EtOH.

5. Place the DNeasy spin column in a 2 ml collection tube (spin columns and collection tubes are provided in the kits by the manufacturer).

6. Place the extracted DNA (or mixture from Step 4) into the spin column, and centrifuge at greater than $6000 \times g$ at room temperature for 1 min. The "flow through" that contains the unwanted cellular material is discarded along with the collection tube.

7. The DNeasy spin column, containing the DNA, is placed in a *new* 2.0 ml collection tube (provided in the kit).

8. Add 500 μl of "Buffer AW1" to the spin column, and centrifuge the column and tube at $6000 \times g$ at room temperature for 1 min.

9. Following centrifugation, the "flow through" again is discarded along with the collection tube, and the spin column placed in a new 2.0 ml collection tube.

10. Pipet 500 μl of "Buffer AW2" into the spin column, and centrifuge the column or tube at full speed for 3 min. This centrifugation step ensures that no residual EtOH is carried over during the following elution.

11. Following centrifugation, the "flow through" is discarded, and the spin column placed in a 1.5 or 2 ml microcentrifuge tube. Pipet 200 μl of "Buffer AE" onto the column, and incubate at room temperature for 1 min.

12. The spin column and microcentrifuge tube are then centrifuged at $6000 \times g$ for 1 min to elute the DNA.

Results

Before any analysis proceeds, it is important to determine the success of this concentration procedure. It is important to determine the quality and quantity of DNA present. It is also important to determine if any degradation of the DNA has occurred. The answers to these questions as well as guidelines for interpreting your results are described in Exercise 4.

Precipitation of DNA Using Ethanol

Analyst: _____   Lab Number: _____

Date: _____

Item Number and/or Description	Extraction Method	Purification Method	Item Number and/or Description	Extraction Method	Purification Method

Comments:

_____
_____
_____
_____

Qiagen Extraction/Purification Method
Manual Spin Columns

Analyst: _____   Date: _____

System: _____   Gel Number: _____

Reagent	Lot/Source
1 X PBS	
Protese	
AL Buffer	/Qiagen
95% EtOH	
AW1 Buffer	/Qiagen
AW2 Buffer	/Qiagen
AE Buffer	/Qiagen

Lysate Transfer Witness: _____

Elution Tube Transfer Witness:_____

Questions

1. What is the purpose of concentrating the extracted DNA?

2. Two different concentration techniques were discussed: ethanol precipitation and column filtration. What are the advantages and disadvantages when using these techniques to concentrate DNA?

5

Exercise 3

Microcon Concentration and Purification of Extracted DNA

Introduction

After DNA extraction, centrifugal filter devices, such as the Microcon purification procedure, can serve as powerful tools in DNA concentration and desalting procedures. Ultrafiltration (UF) is a pressure-driven, convective process that uses semipermeable membranes to separate DNA by molecular size and shape. Ultrafiltration is highly efficient, allowing for concentration and purification at the same time. Unlike the use of chemical precipitation methodologies (as in Exercises 1 and 2 using ethanol or phenol/chloroform), there is no phase change or possible degradation of the DNA with UF. Ultrafiltration routinely concentrates DNA, without the use of co-precipitants, in a short time period with 99% recovery of the starting material. Centrifugal concentrator devices are ideal for separating high and low molecular weight DNA molecules.

The Microcon purification procedure is often used when the biological sample that was extracted was deposited on a substrate (e.g., denim or velvet) known to inhibit DNA amplification or the polymerase chain reaction (see Exercise 6) due to the substrate releasing excessive amounts of dye during the extraction process. In this exercise, the DNA sample is concentrated, then diluted to the original volume with the desired buffer and concentrated again, thus "washing out" inhibitors or the original solvent.

Objective

The purpose of this exercise is to concentrate and remove unwanted components from isolated DNA using ultrafiltration and the Microcon purification procedure.

Equipment and Material

1. Microcon 100 Concentrator Assembly (Millipore Corp.)
2. Adjustable-volume digital micropipets (2–200 µl range)
3. Aerosol-resistant pipet tips
4. 1X TE buffer
5. 1.0 or 1.5 ml microcentrifuge tubes
6. Sterile H_2O
7. Transfer pipets
8. Disposable gloves
9. Phenol:$CHCl_3$:isoamyl alcohol (25:24:1)
10. Bromophenol blue tracking dye (loading buffer)
11. Microcentrifuge

Procedure

The use of disposable gloves and aerosol-resistant pipet tips is highly recommended to prevent cross-contamination. A helpful organizational sheet is provided at the end of the exercise.

A. Concentrating the DNA Using a Microcon Concentrator

1. Use a Microcon concentrator that is adequate for your DNA size (e.g., use the Microcon 50 concentrator for minute biological stains or materials). Add 500 µl of prewarmed (room temperature) phenol:$CHCl_3$:isoamyl alcohol to each tube containing the DNA.
2. Cap the tube, and mix thoroughly by hand for 2–3 sec or until the solution has a milky appearance.
3. Centrifuge the tube(s) for 3 min in a microcentrifuge at greater than 6000 × g to separate the two phases.
4. Insert a labeled Microcon 100 concentrator into a labeled collection vial. Add 500 µl of sterile H_2O to the concentrator. Using a transfer pipet, transfer the aqueous or top phase containing the DNA (from Step 3, above) to the Microcon concentrator. Place the cap from the collection vial on the concentrator.
5. Centrifuge the Microcon assembly in a microcentrifuge for 10–30 min at approximately 5000 rpm (× g or relative centrifugal force [RCF] determined by the rotor used) until the volume is reduced.
6. After centrifugation, remove the concentrator "unit" from the Microcon assembly, and discard the fluid from the filtrate cup. Return the concentrator to the top of the filtrate cup.
7. Remove the cap from the concentrator, and add 200 µl of sterile H_2O. Replace the cap, and centrifuge at 5000 rpm (× g or RCF determined by the rotor used) for 10–30 min until the volume is reduced.
8. Remove the cap from the concentrator, and add 30 µl of 1X TE buffer.
9. Remove the concentrator from the filtrate cup, and carefully invert the concentrator onto the retentate cup. Discard the filtrate cup.
10. Centrifuge the Microcon assembly (the retentate cup end first) at 5000 rpm (× g or RCF determined by the rotor used) for 5 min.

Optional:

11. Remove 4 µl of the sample, and place in a separate 1.0 ml tube. Add 2 µl loading buffer to the sample (see Exercise 4).

12. Run the sample on a test agarose gel for quantitation (see Exercise 4).

Results

Before any analysis proceeds, it is important to determine the success of your extraction, purification, and concentration of the DNA. It is important to determine the quality and quantity of DNA present. It is also important to determine if any degradation of the DNA has occurred. The answers to these questions as well as guidelines for interpreting your results are described in Exercise 4.

Microcon Concentration and Purification of Extracted DNA

Analyst: _____   Lab Number: _____

Date: _____

Item Number and/or Description	Extraction Method	Purification Method	Item Number and/or Description	Extraction Method	Purification Method

Comments:

_____
_____
_____
_____

Questions

1. What are the advantages of the Microcon Concentrator Assembly over the previous extraction processes described?

2. What are the advantages over the two techniques (ethanol precipitation and column filtration) described previously to concentrate extracted DNA?

Chapter 6

Exercise 4
Assessing the Quality and Quantity of Isolated DNA

Introduction

To determine the quality and quantity of extracted or isolated DNA recovered from a sample (known or evidentiary), preliminary tests are conducted. Two tests are often used to assess the quantity (how much DNA is present) or quality (how much, if any, degradation has occurred) of the DNA.

In the first test, a miniature-agarose gel or yield gel is used to estimate both the quality and quantity of DNA recovered from each sample. A yield gel is prepared, and a small portion of each DNA sample is "loaded" into separate wells of the gel. The DNA is analyzed by agarose gel electrophoresis, stained to visualize the DNA by UV illumination, and photographed (or the images are generated using computer software). Documentation for each gel is maintained and indicates the DNA samples that have been included on a particular yield gel, along with the appropriate controls (i.e., visual marker, HindIII-digested lambda DNA; human K562 DNA control or intact lambda DNA for calibration or quantity determination; and undigested K562 DNA). Large, intact, and undegraded DNA will appear as a compact band near the origin of the gel, similar to the standards in the adjacent lanes. Degraded DNA will form a smear and will migrate further through the gel depending on the various sizes of the DNA fragments. Degraded DNA will also be observed following hybridization and autoradiography and/or chemiluminography (Figures 2 and 3). Extremely degraded DNA may not be visible because these smaller fragments will migrate toward the end of the gel. The quantity of the DNA in question can be compared to DNA standards of known quantity that have been run in adjacent lanes.

The second method consists of the slot or dot blot technique, which is used to determine only the quantity of the DNA recovered from a sample. A small portion of the sample DNA in question is applied to a membrane along with a set of standard samples of known quantity. After the samples have been "fixed" to the membrane, a human DNA probe is added and allowed to hybridize to the fixed DNA. The DNA probe used in this instance is tagged or labeled (e.g., with an enzyme or fluorescent dye) for easy detection of the DNA (Figure 4). The slot or dot blot technique does not provide any information on the quality or the level of degradation of the DNA.

FIGURE 2

Restriction Fragment Length Polymorphism (RFLP) Lumigraph (see Exercise 9) Demonstrating DNA Degradation. **Note:** Lanes 1, 5, and 10 contain molecular ladders. Lanes 2 and 9 serve as positive controls and contain a known reference sample. Lane 3 contains degraded DNA from the nonsperm fraction from the victim. Lane 6 contains the suspect's reference sample. Lanes 4 and 8 are blank. Lane 7 contains the sperm fraction from the sexual assault sample. The suspect is a heterozygote (2 DNA fragments) at this locus because two (2) DNA fragments were observed. However, only one (1) faint DNA fragment (the lower molecular weight fragment) was observed at this locus from the sexual assault sample, which would lead the DNA analyst to possibly render an inconclusive result.

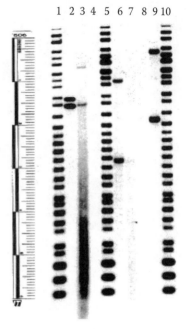

FIGURE 3

Restriction Fragment Length Polymorphism (RFLP) Lumigraph (see Exercise 9) Demonstrating DNA Degradation. **Note:** Lanes 1, 5, and 10 contain molecular ladders. Lanes 2 and 9 serve as positive controls and contain a known reference sample. Lane 3 contains degraded DNA from the nonsperm fraction from the victim. However, two (2) DNA fragments are observed. Lane 6 contains the suspect's reference sample. Lanes 4 and 8 are blank. Lane 7 contains the sperm fraction from the sexual assault sample. The suspect's DNA profile is observed as well as the victim's profile in this sexual assault sample. It should be noted that the "carryover" is evident from the nonsperm fraction into the sperm fraction. Also evident in the sperm fraction are DNA fragments that align with the suspect's two-band profile.

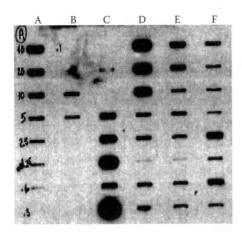

FIGURE 4

Slot Blot Used to Determine the Quantity of Human DNA recovered from a Sample. **Note:** The first column or lane (A) contains the quantitation standards in decreasing order from top to bottom (40, 20, 10, 5, 2.5, 1.25, 0.6, and 0.3 ng DNA). The remaining lanes contain "responses" from the evidence (Lanes B and C) and reference samples (Lanes D and E) as well as from the positive (Lane F) and negative (bottom of Lane B) controls. The amount of DNA contained in a sample (evidentiary or reference) is obtained by comparing the response to the standards.

Objective

To determine the quality and quantity of DNA isolated from a known sample, from an evidentiary sample, or from a human cell line (e.g., HepG, HeLa, and K562) using agarose gel electrophoresis.

Equipment and Material

1. Agarose
2. Ethidium bromide or coomassie blue
3. Ice in buckets
4. TBE buffer
5. K562 DNA
6. 125 ml Erlenmeyer flask
7. Adjustable-volume digital micropipets (2–200 µl range)
8. 1.0 or 1.5 ml microcentrifuge tubes
9. Aerosol-resistant pipet tips
10. Bromophenol blue tracking dye
11. TAE buffer
12. Disposable gloves
13. Power pack or power supply
14. Microwave or hot plate
15. Incubator or water bath at 56°C
16. Electrophoresis systems (gel tray or combs)
17. UV transilluminator
18. Polaroid camera with film

Procedure

The use of disposable gloves and aerosol-resistant pipet tips is highly recommended to prevent cross-contamination. Helpful organizational sheets are provided at the end of the exercise to record your data and observations.

1. Remove 4 µl of extracted and/or resolubilized DNA, and combine with 2 µl tracking dye. This may be done in a centrifuge tube or in a microtiter plate.

2. Preparation of 5% agarose test gel:

    A. Add 1.25 g of DNA Typing Grade Agarose in 25 ml of TAE buffer containing ethidium bromide (EB) at a ratio of 10 µl EB/100 ml TAE buffer.

Note: Some laboratories will stain the gel with ethidium bromide after electrophoresis instead of adding ethidium bromide to the TAE containing the agarose or the reservoir buffer.

    B. Heat the solution in a microwave on high for about 40 sec, swirling the flask by hand every 10–15 sec, or briefly bring to a boil to dissolve the agarose using a hot plate.

    C. Cool the liquid agarose to about 56°C.

    D. While the liquid agarose is cooling, prepare the gel casting tray or gel form according to the manufacturer's guidelines. When cool, pour the agarose into the gel form. Use either one or two 14-well combs.

Note: The number of wells (or teeth) in a comb will vary according to the manufacturer.

    E. Let the liquid agarose "stand" or cool in the gel tray for 10 min to solidify.

    F. Remove gel dams, and pour approximately 175 ml of the TAE–EB buffer into an electrophoresis gel tank.

    G. Remove the comb(s).

Note: If the gel is not ready to be loaded 10 min after pouring, add the buffer anyway so that the gel will not dry out.

3. The DNA sample, mixed with loading buffer (tracking dye), is pipetted into the well with the gel submerged. The final sample volume should contain 10% tracking dye, and the total volume should not exceed 20 µl. Be careful not to push the pipet tip through the bottom of the well in the gel.

4. Include on your gel the following K562 DNA Standards:
       500 ng/4 µl
       250 ng/4 µl
       125 ng/4 µl
       63 ng/4 µl
       31 ng/4 µl
       15 ng/4 µl

Note: Intact lambda DNA can be used in quantities ranging from 300 ng to 10 ng.

5. Set the voltage (100 volts), and "RUN" the samples until the bromophenol blue tracking dye has moved 1–2 cm from the origin (i.e., well) or until the dye front is approximately 2 cm from the end of the gel. This should take less than 20 min.

6. Remove the gel from the electrophoresis tank. Stain the gel in SYBR Green or ethidium bromide (if not added previously to the agarose gel or buffer reservoir).

7. Examine the gel on an ultraviolet (UV) transilluminator. Take a photograph of your gel. Intact DNA will move as a band not far from the origin. A smear from the origin to or past the dye front indicates that the DNA has been fragmented and may not be suitable for further use.

Warning: Avoid excessive exposure to the UV light. Always wear a full face shield when working with the transilluminator.

Yield Gel

Analyst: _____ Lab Number: _____

Date: _____

Well No.	Sample	DNA (ng)
1	Visual marker	
2	Standard	500
3	Standard	250
4	Standard	125
5	Standard	63
6	Standard	31
7	Standard	15
8	Sample	
9	"	
10	"	
11	"	
12	"	
13	"	
14	"	
15	"	
16	"	
17	"	
18	"	
19	"	
20	"	

Reagents	Lot Number	Source
Agarose		
1X TAE (Gel Buffer)		
1X TAE (Tank Buffer)		
Loading Buffer		
Ethidium Bromide		
Visual Marker		
500 ng standard		
250 ng standard		
125 ng standard		
63 ng standard		
31 ng standard		
15 ng standard		

Gel Electrophoresis

Time on:	Voltage:	mAMPs:
Time off:	Voltage	mAMPs:

Gel Prepared By: _____   Date: _____

Tape Your Gel Photo Here

Lane #1 _____

Lane #2 _____

Lane #3 _____

Lane #4 _____

Lane #5 _____

Lane #6 _____

Lane #7 _____

Lane #8 _____

Lane #9 _____

Lane #10 _____

Lane #11 _____

Lane #12 _____

Lane #13 _____

Lane #14 _____

Lane #15 _____

Lane #16 _____

Lane #17 _____

Lane #18 _____

Lane #19 _____

Lane #20 _____

Yield Gel

Analyst: _____   Lab Number: _____

Date: _____

Well No.	Sample	DNA (ng)
1		
2		
3		
4		
5		
6		
7		
8		
9		
10		
11		
12		
13		
14		
15		
16		
17		
18		
19		
20		

Results

Interpreting Your Results

1. While under UV illumination, take a photograph of the gel and attach it to the Reporting Form.

2. From the photograph, estimate the quality of the DNA in the test specimens by comparison to the uncut K562 DNA standards.

   - Intact DNA: Large, intact, and undegraded DNA will appear as a compact band near the origin of the gel, similar to the standards in the adjacent lanes.

   - Degraded DNA: Degraded DNA will form a smear along the lane and will migrate further through the gel depending on the various sizes of the DNA fragments. Degraded DNA will also be observed following hybridization and autoradiography and/or chemiluminography (see Figures 2 and 3). Extremely degraded DNA may not be visible because these smaller fragments will migrate toward the end of the gel.

3. Record the quality of DNA (i.e., intact versus degraded) in the "Results" section, using the "Sample" column in the "Yield Gel" form.

4. From the photograph, estimate the quantity of DNA in test specimens. The quantity of the DNA in question can be compared to DNA standards of known quantity that have been run in adjacent lanes.

   - A single estimate should be made (not a range), and the quantity of total DNA remaining per 4 µl and 32 µl of the sample should be recorded on the worksheet. The estimation is multiplied by 8 to obtain the total quantity of DNA in the remaining 32 µl of the sample. Record the estimated quantity of DNA in the "Results" section using the "Yield Gel" form.

Questions

1. Once the DNA has been isolated and before any analysis can proceed, it is imperative to determine the quality and quantity of DNA present. Why is it important to determine the quantity and quality of DNA in a sample?

2. How would a DNA analyst determine, in one (1) experiment, the quantity and quality of DNA from a given sample?

3. If the evidentiary sample to be analyzed was degraded, what are the many "forms" of DNA that would be expected following gel electrophoresis?

Exercise 5

DNA Analysis Using Restriction Fragment Length Polymorphisms (RFLP)

Introduction

The first widespread use of DNA tests involved restriction fragment length polymorphism (RFLP) analysis, a test designed to detect variations in the DNA from different individuals. In the RFLP method, DNA is isolated from a biological specimen (e.g., blood, semen, and vaginal swabs) and digested or cut by enzymes called restriction endonucleases (or RENs). The resulting DNA or restriction fragments are separated by size into discrete bands by gel electrophoresis. Following electrophoresis, the separated restriction fragments are transferred onto a membrane by Southern "blotting," and identified using probes (known DNA sequences that are "tagged" with a chemical tracer). The resulting DNA profile is visualized by exposing the membrane to a piece of X-ray film, which allows the scientist to determine which specific fragments the probe identified among the thousands in a sample of human DNA. A "match" is made when similar DNA profiles are observed between an evidentiary sample and a suspect's DNA or known reference sample. A determination is then made as to the probability that a person selected at random from a given population would match the evidence sample as well as the suspect. The entire analysis may require several weeks to complete.

Objective

In this exercise, you will gain experience with RFLP, the first test adapted for forensic DNA analysis. During this exercise, you will learn how to use restriction enzymes to digest or cut DNA and how to separate the resulting DNA fragments by agarose gel electrophoresis. You will also learn how to transfer DNA fragments onto a membrane (for support) by Southern blotting to allow for the detection of allelic patterns following hybridization.

Equipment and Material

1.   1.0, 1.5, and 2.0 ml Eppendorf or microcentrifuge tubes
2.   HaeIII restriction enzyme
3.   Adjustable-volume digital micropipets (2–200 μl range)
4.   Aerosol-resistant pipet tips
5.   HindIII-digested lambda DNA
6.   HaeIII-digested K562 DNA
7.   KpnI-digested Adenovirus DNA
8.   Molecular weight markers (526 to 22,621 bp)
9.   Herring sperm DNA (10 mg/ml)
10. Ethidium bromide or coomassie blue
11. Bromophenol blue tracking dye
12. Kimwipes™
13. Neutralization solution
14. 125 ml Erlenmeyer flask
15. TBE buffer
16. Agarose
17. 1X TAE buffer
18. Sterile distilled water
19. Ice in buckets
20. 0.5X Wash I solution
21. 1X Wash I solution
22. 1X Wash I solution concentrate
23. Labeled DNA probes
24. Plastic box or tray
25. Glass baking dish
26. 30 or 50 ml conical tube
27. Nylon membrane
28. Denaturation solution
29. Disposable gloves
30. 20% SDS
31. 20X SSPE
32. 10X SSC
33. 50% PEG
34. TE buffer
35. Thin sponges

36. Whatman 3MM chromatography paper

37. Kodak X-Omat RP film and cassette

38. Blotting pads or paper towels

39. Glass or plastic plate

40. Glad or Saran wrap

41. Orbital shaker or rotator

42. Vortex mixer

43. Microwave or hot plate

44. Electrophoresis systems (gel tray or combs)

45. Incubator or water bath at 37°C, 55°C, and 65°C

46. UV transilluminator

47. Microcentrifuge

48. Vacuum oven

49. Power pack or supply

50. Polaroid camera with film

Procedure

The use of disposable gloves and aerosol-resistant pipet tips is highly recommended to prevent cross-contamination. Helpful organizational sheets and semilog graph paper are provided at the end of the exercise.

A. Digesting DNA with Restriction Endonucleases (RENs)

Restriction endonucleases are enzymes that recognize specific sequences within double-stranded DNA. Hundreds of different RENs have been isolated, and most recognize different sequences. To ensure uniformity and consistency in the forensic community, the most often used REN for forensic casework is HaeIII. As with all RENs, HaeIII has a set of optimal reaction conditions specified on the product insert or information sheet provided by the manufacturer. The major variables in the restriction digests are the temperature of incubation and the composition of the reaction buffer. Restriction digests or reactions typically contain approximately 1 µg of DNA in a volume of 20 µl or less. Once the concentration of DNA in the sample(s) has been determined (see Exercise 4), the volume of DNA sample to be added to the reaction is adjusted accordingly.

1. If the quantity of DNA in the sample exceeds 1000 ng, dilute the sample in a microcentrifuge tube and bring the final volume up with sterile water to: a) a minimum of 18 µl to, b) a maximum of 32 µl.

2. To each sample containing DNA, add: a) 2.0 µl of 10X digestion buffer (supplied with enzyme) or, b) 4.0 µl of 10X digestion buffer.

3. Add 1 unit of the restriction enzyme (HaeIII), mix by tapping the tube, and centrifuge the microcentrifuge tube containing the sample(s) for 2 sec.

Note: One unit of enzyme is defined as the amount required to digest 1 µg of DNA to completion in one (1) hr in the recommended buffer and at the recommended temperature, usually 37°C. The volume of restriction enzyme added should never be more than 10% of the final digestion volume.

Note: Always keep HaeIII or any restriction enzyme on ice.

4. Incubate the reaction tube at 37°C for 1 hr.

5. Centrifuge the reaction tube for 5 sec.

B. Digest Gel to Measure Completeness of Restriction Digestion

The purpose of the digest gel is to assess the completeness of the HaeIII digestion of the specimen DNA before proceeding to an RFLP analytical gel (see below). Controls include HindIII-digested lambda DNA and approximately 200 ng (3 µl) of HaeIII-digested K562 DNA.

1. To determine that the reaction is complete and that the sample DNA has been completely digested, remove 4 µl of the HaeIII-digested DNA, and combine with 2 µl loading buffer (tracking dye). This may be done in a microcentrifuge tube.

2. Run the digested sample(s) on an agarose test gel to determine if restriction digestion is complete (see Exercise 4 for details on preparing an agarose test gel).

3. Completely digested DNA will be present on this gel as a uniform smear composed of different-sized fragments "running" from the origin down the length of the lane to the dye front (Figure 5, Lane 4). Proceed to the next step ("C. Resolution of DNA Fragments on an Analytical Gel") if the samples are completely digested. Uncut DNA will form a compact, fluorescent large band near the origin of the gel (Figure 5, Lane 3). Partially digested DNA will appear as multiple high and low molecular weight DNA bands (Figure 5, Lane 5).

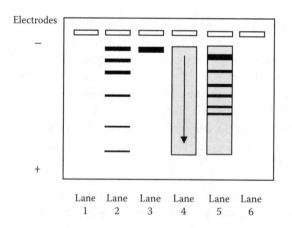

FIGURE 5

Digest Gel Demonstrating Completely and Incompletely Digested DNA Samples. *Note*: Lanes 1 and 6 are blank (no DNA); Lane 2 contains the molecular weight markers; Lane 3 contains a single, high molecular weight band demonstrating undigested DNA; Lane 4 contains completely digested DNA illustrating a uniform "smear" of high to low molecular weight DNA fragments (also refer to Figures 2 and 3, Lane 3); and Lane 5 contains partially digested DNA showing evidence of a substantial amount of high molecular weight DNA.

C. Resolution of DNA Fragments on an Analytical Gel

DNA fragments generated by REN digestion are separated by conventional gel electrophoresis using a larger agarose gel format that is often referred to as an analytical gel. The analytical gel is designed to separate the REN-digested DNA (i.e., the HaeIII-digested DNA isolated from the different specimens) based on size. The analytical gels contain several controls: KpnI-digested Adenovirus DNA (nine fragments ranging in size from 1086 to 771 bp), molecular weight markers (30 viral DNA fragments ranging in size from 526 to 22,621 bp), and HaeIII-digested K562 DNA (human DNA control). The analytical gels are generally composed of 1% agarose in 1X TAE buffer. The gel dimensions are 11 × 14 × 0.65 cm (usually less than 100 ml total volume is needed).

1. Prepare the analytical gel:

    A. For each analytical gel, prepare 1 liter of 1X TAE buffer (50 ml of 20X TAE combined with 950 ml of distilled H_2O).

    B. To a 125 ml Erlenmeyer flask, add 1.0 g of agarose into a flask. Add 100 ml of 1X TAE buffer containing ethidium bromide (EB; 10 μl EB/100 ml of TAE buffer).

Optional: Ethidium bromide (20 mg/ml) can be added directly to the gel at this stage, or the gel can be stained in EB following electrophoresis.

    C. Microwave on high for about 40 sec. Swirl the flask and continue to microwave the agarose solution on high for another 40 sec, *or* bring to boil to dissolve agarose using a hot plate.

    D. Prepare the gel-casting tray (according to the manufacturer's guidelines) so that the open ends are sealed or secured and the tray is level.

    E. Place a comb into the gel-casting tray with the flat edge of the comb nearest you. The number of samples to be analyzed will help determine the number of "teeth," or wells, on the comb needed.

    F. Allow the agarose to cool to 50–60°C prior to casting. After pouring the agarose into the gel tray, let the solution "cool" for at least 15 min to solidify.

2. Pour the remaining 900 ml of the 1X TAE buffer into the gel tank.

3. Remove the "ends" from the gel-casting tray, and place the tray into the tank with the well comb nearest you. The TAE buffer should cover the gel to a depth of at least 0.5 cm. Remove the comb from the gel.

4. Load the samples into the analytical gel:

    A. Aliquot 12 μl of the KpnI-digested Adenovirus DNA (analytical visual marker) into a 1.5 ml microcentrifuge tube.

    B. To prepare the molecular weight (MW) marker, determine how many marker lanes you need. Follow the manufacturer's directions to prepare the MW markers.

    C. To prepare sample DNAs, add 14 μl of HaeIII-digested DNA (see Step A, above) and 4 μl of loading buffer to a microcentrifuge tube. Mix, and centrifuge for 2 sec. Repeat for all samples to be analyzed. If less than 14 μl of the digested DNA is used, add TE to bring the volume to 14 μl.

    D. Carefully pipet each sample into the appropriate well as follows:

Well	Sample
1	Blank
2	Analytical visual marker (10 µl)
3	DNA analysis marker (10 µl)
4	HaeIII-digested K562 DNA (18 µl)
5	Test sample
6	Test sample
7	DNA analysis marker (10 µl)
8	Test sample
9	Test sample
10	DNA analysis marker (10 µl)
11	Test sample
12	Known male standard*
13	DNA analysis marker (10 µl)
14	Blank

Note: The first and last lanes or wells should be left empty unless they are needed.

* A known male bloodstain is extracted and run with other test samples with casework.

5. Connect the electrodes from the power pack or supply to the gel electrophoresis apparatus. An electric field is then applied to the agarose gel by setting the voltage to 100 volts. Because the DNA molecules are negatively charged, the fragments will migrate toward the positive (+) electrode (red). Consequently, the wells containing the samples are located at the opposite end of the gel, or close to the negative (–) electrode (black).

6. After electrophoresis (determined when the "dye front" is approximately 2 cm from the end of the gel), examine the gel on the UV transilluminator to evaluate migration. Under UV illumination, the ethidium bromide (which has bound to the DNA) fluorescence should appear as a smear without the appearance of any bands.

Warning: Avoid excessive exposure to the UV light. Always wear a full face shield when working with the transilluminator.

D. Southern Blotting DNA from Agarose Gels

To detect specific polymorphic DNA fragments, the DNA must be transferred from the agarose or analytical gel to a solid support. This transfer is accomplished by a method described by Southern (1975) and is referred to as the "Southern blot." The polymorphic fragments, separated by agarose gel electrophoresis based on size, are denatured (i.e., separated into single strands) using an alkaline solution, transferred to a solid support (e.g., a nylon membrane is preferred over nitrocellulose), immobilized, and detected with DNA probes through a process called hybridization. The DNA profile or pattern is captured on film (e.g., autoradiography or chemiluminescence) for analysis and long-term storage (for an overview of the RFLP process, including Southern blotting or the DNA transfer setup, see Figure 6).

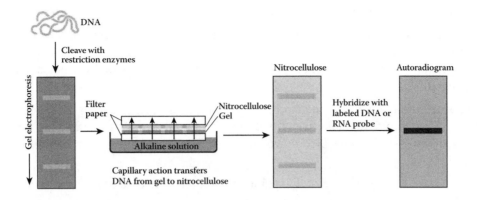

FIGURE 6

Overview of the RFLP Procedure. *Source*: From the University of Strathclyde in Glasgow, http://www.strath.ac.uk/~dfs99109/ BB211/RecombDNAtechlect2.html.

1. Slide the gel from the tray into a plastic box that contains enough denaturation solution to cover the gel. Place the box containing the gel on an orbital shaker, and gently stir for 15 min.

2. While the gel is soaking in denaturation solution, place thin sponges (approximately 5 cm larger than the gel on all sides) in a glass dish or transfer tray, and add 10X SSC until sponges are saturated and solution level is just above the bottom sponge.

3. In a separate tray containing 10X SSC, immerse an 11.5 × 14.5 cm prelabeled (using pencil) nylon membrane. The transfer nylon membrane should be handled at the edges.

Note: Always wear disposable gloves when handling nylon membranes. Oils from your skin will prevent proper wetting of the membrane and subsequent transfer of DNA.

4. Decant the denaturation solution, and rinse the gel in deionized or distilled water for 20 sec. *Gently* shake by hand, and then decant water. Add enough neutralization solution to cover the gel, and place the tray containing the gel on an orbital shaker and shake gently for 15 min. Ensure that the solution covers the gel and that the gel is not adhering to the bottom of the tray.

5. Cut three pieces of Whatman 3MM chromatography paper to the same size of the sponge, saturate with 10X SSC, and place on top of the sponge.

6. Carefully remove the gel from the neutralization solution, and place on the chromatography paper with the underside of the gel facing up. Remove any trapped air bubbles between the paper and the gel by rolling a pipet over the surface of the gel.

7. Place the transfer nylon membrane (from Step 3, above) on the surface of the gel, and remove any air bubbles as in the preceding step.

8. Cover the membrane with a piece of Whatman 3MM chromatography paper that has been cut to the size of the gel (11 × 14 cm) and presoaked with 10X SSC. Remove any air bubbles as before with a pipet.

9. Place 9 dry blotting pads or paper towels cut to the size of the gel, and place on top of the Whatman 3MM chromatography paper.

10. Place a glass or plastic plate on top of the blot pads or paper towels.

11. Place a weight on top of the glass or plastic plate to hold the assembly in place.

12. Allow the transfer to proceed at room temperature for at least 2 hr or until all blot pads are saturated. Transfer should not exceed 6 hr. Due to the limited volume of 10X SSC, check periodically and add 10X SSC as needed. **Do not allow the Southern blot to dry out.**

13. Remove the blotting pads or paper towels and the Whatman 3MM chromatography paper.

14. Grasp the membrane at the right corner (origin end) with your gloved hand, turn it over, place in a plastic tray, and wash once with 0.2 M Tris, pH 7.5, and 2X SSC for 15 min with gentle shaking on a rotator. Place the membrane on a sheet of Whatman 3MM chromatography paper, and allow to dry at room temperature.

15. To immobilize the DNA to the membrane, either place the membrane in an envelope made from Whatman 3MM chromatography paper and place in an 80°C vacuum oven for 30 min, or irradiate the membrane (DNA side face down) for 1 to 5 min using a UV transilluminator (254 nm wavelength).

16. The membrane can be hybridized at this point or stored in a Ziploc plastic bag at room temperature.

E. Hybridization and Detection

Chemiluminescent Probe

A number of methods have been used in forensic laboratories for detecting DNA following electrophoretic separation and Southern blotting. These methods were effective but were expensive and time consuming. During the past decade, fluorescence-based detection systems have gained in popularity and are widely used in forensic laboratories due to their ease of use and rapid formats.

With RFLP analysis, the single-stranded DNA probe is tagged with an enzyme and hybridized to the DNA fragments immobilized on the Southern blot. When the Southern blot is exposed to a certain chemical, a discharge of light is emitted from the enzymatically labeled probe that has bound to its complementary strand during hybridization. This effect or process is referred to as chemiluminescence. The light that is emitted from the chemical is captured on X-ray film as dark black bands in the region detected by the probe as well as with computer-aided imaging systems.

1. Prepare 500 ml of 1X Wash I solution per hybridization container and 250 ml of 0.5X Wash I solution. Preheat solutions to 55°C prior to hybridization. Both solutions can be prepared a day in advance and stored at 55°C until they are ready for use, *or* incubate at 55°C approximately 2 hr prior to use.

Note: Preheat Wash I Concentrate at 55°C prior to making dilution to ensure that all solids are dissolved. If concentrate becomes cloudy, leave at room temperature until clear.

2. Add membranes, DNA side up, to the appropriate volume (e.g., 1–4 membranes use 30 ml of prehybridization or hybridization solution, and 5–8 membranes use 60 ml of prehybridization or hybridization solution) of hybridization solution (Wash I Concentrate). Ensure that each membrane is covered with solution prior to adding the next membrane. This "blocking" or prehybridization step may be performed in a sealable bag or a tray.

3. Prehybridize membranes for 20 min in a rotating water bath (60–70 rpm)* at 55°C.

4. During the prehybridization period, transfer the appropriate amount of hybridization solution (see Step 2, above) to a conical tube and heat to 55°C. To the conical tube, add the appropriate amount of chemiluminescent-labeled probe (as described in the manufacturer's product information; between 15 and 30 μl per 30 ml hybridization solution) and the molecular weight marker probe.

5. Remove the hybridization box from the water bath, and decant the prehybridization solution.

* In an orbital shaker; speed control and/or settings may vary between instruments.

6. Add the appropriate volume of labeled probe and molecular weight marker probe to a 50 ml conical bottom centrifuge tube containing the appropriate volume of hybridization solution (see Step 4, above).

7. Vortex the hybridization solution and labeled probes solution briefly, and add to the hybridization container. Rotate the container gently by hand to distribute the hybridization solution, ensuring that each membrane is covered with solution.

8. Hybridize membranes for 30 min in a rotating water bath (60–70 rpm) at 55°C in hybridization solution. To ensure equal distribution of the probe, the membranes should not adhere to each other.

Radioactive Probe

1. Mix the following solutions (hybridization solution) into a hybridization container:
   40.8 ml sterile H_2O
   24.0 ml 50% PEG
   9.0 ml 20X SSPE
   42.0 ml 20% SDS
   (As many as 6 nylon membranes can be added to the hybridization container.)

2. Place 60 ml of the hybridization solution into the hybridization container.

3. Using gloves, place the membrane(s) into the hybridization container one at a time. Make sure that each membrane is saturated thoroughly with the hybridization solution before adding the next membrane to the tube.

4. Place the hybridization container in the 65°C incubator shaker to equilibrate.

5. Fill a beaker with water, and place on a hot plate to boil.

6. For each hybridization, label a 15 ml conical screw-capped centrifuge tube with the name of the probe (e.g., D2S44 and D10S28).

7. Pipet 1.5 ml of herring sperm DNA into each conical screw-capped tube.

8. Pipet the entire contents of the vial of labeled probe[*] into the tube(s) containing the herring sperm DNA. Dispose of the pipet tip and probe vial in an appropriate radioactive waste container. Cap the tube securely.

Note: DNA labeling kits are commercially available through various vendors.

9. Place the conical tube(s) in the boiling water for 5 min.

10. Remove the hybridization container from the incubator shaker, and remove the lid.

11. Lift the hybridization container to one side so that all of the "prehybridization solution" collects in one corner.

12. Carefully pour the contents (or probes) of the conical tube(s) into the corner of the container, and carefully rotate the container to mix the solutions. Dispose of the empty conical tube in a radioactive waste container.

13. Replace the lid on the hybridization container, and incubate at 65°C overnight with constant shaking.

---

[*] A short segment of DNA that are "tagged" with a chemical tracer and used to detect alleles at certain loci.

F. Posthybridization Washes

Chemiluminescent Procedure

1. Decant the hybridization solution from the hybridization container.

2. Rinse the membranes in the following washes in a rotating water bath (i.e., orbital shaker):

Note: The membranes will stick to the bottom of the container. To ensure effective washing, make sure that the membranes are not adhering to each other during the posthybridization washes. Use enough wash solution to cover the membranes and half-fill the container. The speed control and/or settings may vary between shakers.

   A.   15 min in 1X Wash I solution at 55°C at 60–70 rpm

   B.   15 min in 1X Wash I solution at 55°C at 60–70 rpm

   C.   15 min in 0.5X Wash I solution at 55°C at 60–70 rpm

   D.   5 min in 1X Final Wash solution at room temperature with gentle agitation

3. Place membrane(s) DNA side up on a clean sheet of Whatman 3MM chromatography paper to remove excess solution. **Do not blot the membrane(s)**. After 5–10 min, place the air-dried membranes in a plastic container with the appropriate volume of LumiPhos Plus (see below). Make sure each membrane is covered with LumiPhos Plus before adding the next membrane. Place the container on a rocking or shaking platform, and gently shake for 5 min.

Membrane Number and Appropriate LumiPhos Plus Volumes	
Number of Membranes	LumiPhos Plus Volume
1–2	15 ml
3–4	20 ml
5–6	25 ml
7–8	30 ml

4. Carefully remove each membrane from the LumiPhos Plus solution using blunt-end forceps. To remove excess LumiPhos Plus, "drag" the membrane along the side of the container.

5. Place each membrane in a plastic folder, wipe folder with a Kimwipe™ to press out air bubbles, and heat-seal folder.

6. Trim the excess plastic close to the outer edge of the heat seal. Wipe the edges with a Kimwipe™ to remove any excess LumiPhos Plus. The plastic folders can accommodate two membranes per folder.

Radioactive Procedure

1. Decant the hybridization solution from the hybridization container slowly into a radioactive waste container. Wipe the corner of the container with a Kimwipe™, and discard the waste into a radioactive waste container.

2. Rinse the membranes in the following washes in a rotating water bath (i.e., orbital shaker):

Note: The membranes will stick to the bottom of the container. To ensure effective washing, make sure that the membranes are not adhering to each other during the post-hybridization washes. Use enough wash solution to cover the membranes and half-fill the container.

A. 15 min in 2X SSC + 0.1% SDS at room temperature with slow "rocking"

B. 15 min in 2X SSC + 0.1% SDS at room temperature with slow "rocking"

C. The final *high-stringency* wash conditions (0.1X SSC + 0.1% SDS and 65°C) will vary according to the probe used (see below):

Probe	Number of Washes	Length of Each Wash
D2S44	1	10 min
D17S79	1	10 min
D1S7	2	30 min
D4S139	2	30 min
D1OS28	2	30 min
D5S11O	2	30 min

Note: For the final stringency washes, the 0.1X SSC + 0.1% SDS solution must be preheated to 65°C before being added to the membrane(s).

3. Carefully remove each membrane from the final wash solution using blunt-end forceps, and place on Whatman 3MM chromatography paper.

G. Lumigraphy

1. The plastic folders containing the membrane(s) are stored overnight *in the dark* at room temperature to allow for maximum light output.

2. In the darkroom under red light illumination, place the membrane-containing folders, DNA side down, onto Kodak X-Omat RP film. Tape the membrane packets to the film. Record the locations of the membrane(s) by writing directly on the film with a ballpoint pen. Place a second sheet of film onto the back of the membrane, and place in a cassette. Close the cassette, and keep at room temperature.

3. The two sheets of Kodak X-Omat RP film can be developed after an exposure period has been determined for the film in use. This exposure period can range from 15 min to 60 min for the back film, and 30 min to 2 hr for the front film. Film exposure times in excess of 2 hr will, in general, not increase the band or signal intensity but will increase background noise or signal.

H. Autoradiography

1. Wrap the damp membrane(s) in Glad wrap. The wrap prevents contamination of holders and prevents the membrane from sticking to the film.

2. In the darkroom under red light illumination, place the membranes onto the X-ray film. Tape the membrane(s) securely on the film. Record the locations of the membrane(s) by writing directly on the film with a ballpoint pen. Place another sheet of film onto the back of the membranes. Close the cassette, and place at −80°C.

3. The X-ray film on the back side of the membrane(s) can be removed after a short exposure period, usually within a few hrs or as long as a few days. The back sheet of film, when developed, is used as a guide for determining the length of exposure time for the front film.

Results

Digest Gel

Tape Your Gel Photo Here

Lane #1 _____

Lane #2 _____

Lane #3 _____

Lane #4 _____

Lane #5 _____

Lane #6 _____

Lane #7 _____

Lane #8 _____

Lane #9 _____

Lane #10 _____

Lane #11 _____

Lane #12 _____

Lane #13 _____

Lane #14 _____

Lane #15 _____

Lane #16 _____

Lane #17 _____

Lane #18 _____

Lane #19 _____

Lane #20 _____

Analytical Gel for RFLP Analysis

Analyst: _____   Lab number: _____

Date: _____

Well No.	Sample	DNA (ng)
1	Blank	
2	HindIII-digested lambda DNA (18 µl)	
3	HaeIII-digested K562 DNA (18 µl)	
4	Test sample	
5	Test sample	
6	Test sample	
7	Test sample	
8	DNA analysis marker (10 µl)	
9	Test sample	
10	Test sample	
11	Test sample	
12	Test sample	
13	DNA analysis marker (10 µl)	
14	Test sample	
15	Test sample	
16	Test sample	
17	Test sample	
18	Known male standard	
19	DNA analysis marker (10 µl)	
20	Blank	

Reagents	Lot No.	Source
Agarose		
1X TAE (gel buffer)		
1X TAE (tank buffer)		
Loading buffer		
Ethidium bromide		
Visual marker		

Analytical Gel for RFLP Analysis

Analyst: _____   Lab number: _____

Date: _____

Well No.	Sample	DNA (ng)
1		
2		
3		
4		
5		
6		
7		
8		
9		
10		
11		
12		
13		
14		
15		
16		
17		
18		
19		
20		

Quantitative Analysis of DNA Fragment Sizes

To make a comparison between the test samples (or crime scene samples) and the known reference samples, other than a visual comparison, a quantitative measurement of the DNA fragments observed following hybridization and visualization is needed. Determining the molecular weight of each DNA fragment is performed as follows:

1. Using a ruler, measure the distance that each DNA fragment (i.e., the molecular weight standards, the known reference samples, and the test samples) or band has migrated from the bottom of the well (the origin) to the center of each DNA fragment.

2. To determine the molecular weight or size of each DNA fragment, a standard curve is created using the distance (x-axis) and fragment size (y-axis) data from the molecular weight markers (HindIII-digested lambda DNA and/or the DNA analysis markers). Using both the linear and semilog paper (provided at the end of this exercise), plot the distance that each DNA band has migrated (in mm) versus the size (in base pairs) for each band of the molecular weight markers. Record the molecular weight of each DNA fragment in either table below. After each data point has been "plotted" on the graph paper, use a ruler and draw the "best-fit" line through the data points. Extend the line through the entire graph paper.

3. Compare the results of your graphs. Decide which graph should be used to estimate the size of the DNA fragments observed from the test samples.

4. To estimate the size of a DNA fragment from the test sample, determine the distance the fragment migrated. Locate the distance migrated on the x-axis of your graph. Read "up" to the standard line, and then follow the graph line over to the y-axis. This point is the approximate size (in base pairs) of the unknown DNA fragment. Record the molecular weight of each DNA fragment in either table below. Repeat this process for each DNA fragment.

5. Compare the sizes of the DNA fragments of the test samples to the reference samples. Is there a match?

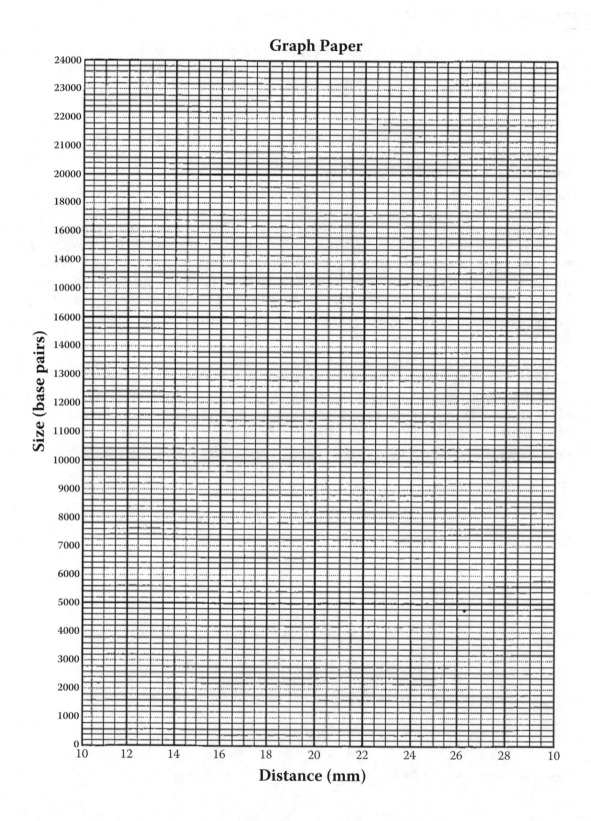

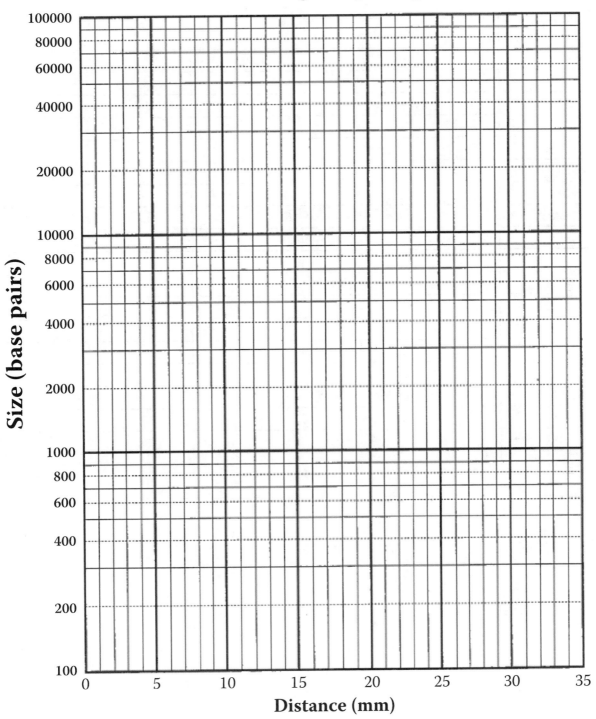

RFLP Typing Results

Sample	Probe/Locus				
1					
2					
3					
4					
5					
6					
7					
8					
9					
10					
11					
12					
13					
14					
15					
16					
17					
18					
19					
20					

RFLP Typing Results

Band	Lambda/Hind III Size Marker		Test Samples (Crime Scene)		Reference Sample		Reference Sample	
	Distance (mm)	Actual Size*	Distance (mm)	Actual Size*	Distance (mm)	Actual Size*	Distance (mm)	Actual Size*
1		23,130						
2		9,416						
3		6,557						
4		4,361						
5		2,322						
6		2,027						

* The actual size of the DNA fragments is an approximation of the molecular weight in base pairs (bp).

RFLP Typing Results

Band	Lambda/Hind III Size Marker		Test Samples (Crime Scene)		Suspect 1		Suspect 2	
	Distance (mm)	Actual Size*	Distance (mm)	Actual Size*	Distance (mm)	Actual Size*	Distance (mm)	Actual Size*
1		23,130						
2		9,416						
3		6,557						
4		4,361						
5		2,322						
6		2,027						

* The actual size of the DNA fragments is an approximation of the molecular weight in base pairs (bp).

Questions

1. Restriction endonucleases are used to cleave the DNA into many fragments. If the sample was not completely digested, what type of restriction pattern would you expect to observe following gel electrophoresis? Assuming that the sample will need to be redigested, what steps would you take to ensure that the sample was completely digested?

2. During gel electrophoresis, an electric field is created with positive and negative poles at the ends of the gel. After the DNA sample is loaded into the wells, to which electric pole would you expect DNA to migrate? Explain your answer.

3. After the DNA samples have been loaded into the sample wells, what size fragments (large versus small) would you expect to move toward the opposite end of the gel the quickest?

4. When looking at the data plotted on the linear and semilog paper, which graph provides the best-fit line that would allow you to estimate the test samples' or known reference samples' size?

Chapter 8

Polymerase Chain Reaction (PCR)

Introduction

Early forensic detection systems relied on the quality and quantity of the DNA sample to be analyzed. The large amount (a "dime-size" stain) of isolated DNA, whether from an evidentiary or a reference sample, had to be relatively fresh or undegraded—essentially, unadulterated—for these detection systems to yield a sufficient profile. For samples considered too minuscule (i.e., low concentration) or determined to be degraded, the polymerase chain reaction (PCR) or amplification process is now performed. The process initially involves the isolation of DNA from a biological specimen followed by a quantitative and qualitative assessment of recovery. The PCR amplification technique is used to produce millions of copies of a specific portion of a targeted chromosome(s) that contains polymorphic DNA selected for forensic and parentage evaluations. The abundance of the PCR product allows for the direct visualization of "blue dots" or bands that represent the allele(s) at specific loci. This evaluation is accomplished by reverse dot blot analysis or gel electrophoresis followed by chemical staining. DNA typing using PCR and gel electrophoresis eliminated the need for critically sensitive DNA probes previously used with the restriction fragment length polymorphism (RFLP) procedure. Analysis time ranges between 24 to 48 hours. Computer-assisted image analysis of test results is helpful but not always necessary because the resulting genetic profiles are routinely interpreted by visual or direct comparison to allele standards at specific loci. Population frequencies are conservative estimates based upon classical population genetic principles.

The PCR methodology also has demonstrated consistency of results from tissue to tissue, and from body fluid to body fluid, within an individual. Therefore, the detection of an allele(s) in an unknown biological stain allows for comparison to reference samples regarded as a possible source or contributor to such biological materials. If the collection of genetic information (DNA profile) associated with an unknown stain is consistent with the results generated from a reference sample, then the possibility that a common genetic source exists for both sets of samples cannot be eliminated. The demonstration of independent Mendelian inheritance, as mentioned earlier, allows for a conservative estimate of the frequency of such a DNA profile occurring among unrelated individuals randomly selected in various racial groups.

Objective

The following exercises (Exercises 6–10) will introduce the student to different PCR-based tests: the Ampli-Type PM/DQA1 system, the D1S80 system, and short tandem repeat (STR) analysis.

Laboratory Setup

Due to the sensitivity of PCR-based tests, certain precautions are necessary to avoid contamination of samples with other sources of DNA. To minimize the potential for laboratory-induced DNA contamination, several aspects of the PCR process should be considered: 1) DNA extraction, 2) PCR setup, and 3) amplified DNA analysis. Each aspect of the PCR process should be separated by time and space. The following section addresses special precautions that must be taken to minimize contamination in the laboratory.

1. The work area for DNA extractions and non-amplified DNA should include dedicated equipment and supplies.

2. DNA extractions and PCR setup should be conducted within self-contained hoods. If hoods are unavailable, use an area of a benchtop that is dedicated for this use only.

3. Use disposable gloves at all times, and change frequently. Prior to leaving the laboratory area, always remove the gloves and wash your hands.

4. DNA extraction of questioned samples (i.e., evidentiary samples) should be performed separately from the extraction of known samples. This will minimize the potential for cross-contamination between samples.

5. Every sample to be analyzed should be properly labeled and recorded. Evidentiary and known reference samples to be analyzed in a forensic laboratory are given a unique identification number that is used throughout the entire analysis.

6. To minimize sample-to-sample contamination, extract samples containing high levels of DNA (e.g., whole blood) separately from samples containing low levels of DNA (e.g., small bloodstains, stamps, and envelopes).

7. Always use sterile solutions or reagents and, whenever possible, sterile disposable supplies (i.e., pipet tips and microcentrifuge tubes).

8. Always change pipet tips between handling each sample even when dispensing reagents.

9. Sterilize reagents, and store as small aliquots to minimize the number of times a given tube of reagent is opened. It is recommended that the small aliquots be retained until typing of the set of samples for which the aliquots were used is complete. Then dispose of the tube containing the reagent.

10. Include reagent blank controls with each set of DNA extractions.

11. Before and after setting up the DNA extractions, clean all work surfaces thoroughly with a 10% solution of bleach. In addition, the use of disposable bench paper will prevent the accumulation of human DNA on permanent work surfaces.

9

Exercise 6

Polymerase Chain Reaction (PCR)–Based Tests: The AmpliType PM/DQA1 System

Introduction

If the evidentiary sample contains an insufficient quantity of DNA or if the DNA is degraded, a polymerase chain reaction (PCR)–based test may be used to obtain a DNA profile. The PCR-based tests generally provide rapid results that can serve as an alternative or as a complement to other DNA testing. The process involves the isolation of DNA from a biological specimen (e.g., blood, semen, saliva, or fingernail clippings) followed by an assessment of DNA quality and quantity. Next, the PCR amplification technique is used to produce millions of copies of a specific portion of a targeted DNA segment (Figure 7). The PCR amplification procedure is similar to a molecular photocopying machine. The amplified PCR products are then separated and identified by the reverse dot blot technology or by gel electrophoresis followed by enzymatic conversion of a colorless substrate or by chemical staining using ethidium bromide or coomassie blue, respectively. The resulting DNA profiles are routinely interpreted by direct comparison to DNA and/or allele standards. Probability calculations are determined based upon classical population genetic principles.

The first PCR-based test that became available to the forensic community was the HLA DQ alpha system (developed by Cetus), now referred to as DQA1. The human leukocyte antigen (HLA) locus was found to contain a significant number of variations that, when applied to human identification, were able to distinguish 28 DQA1 genotypes. However, because only one locus with limited variability was analyzed in this system, the power of discrimination was extremely low. A second commercially available PCR typing kit, called the AmpliType PM kit or Polymarker (the manufacturer-distributor's name has changed from Roche Molecular to Applied Biosystems, Foster City, California), was introduced in the mid-1990s with the ability to type 5 additional loci along with the HLA DQA1 locus. The combination of 6 genetic loci increased the power of discrimination while retaining the advantages of a PCR-based system.

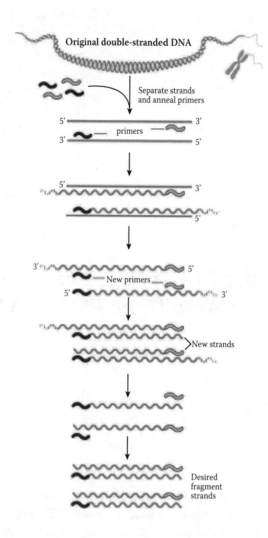

FIGURE 7

Overview of the Polymerase Chain Reaction (PCR). **Note:** A defined region of DNA is copied using the PCR amplification technique to produce millions of copies of the specified region on the DNA strand.

The format for the PM/DQA1 typing process is quite different than the format for RFLP. The PM/ DQA1 test took advantage of the reverse dot blot technology where the DNA probe is bound to a solid substrate (i.e., a nylon membrane) and the amplified PCR products—from known and evidentiary samples—are hybridized to two separate nylon strips containing the immobilized probes for either the DQA1 amplified products or the 5 PM loci. The final product(s) or the positive responses are visualized upon the enzymatic conversion of a colorless substrate to a blue-colored precipitate (Figure 8). The pattern of "blue dots" corresponds to the alleles detected in the samples tested.

Objective

In the next series of exercises, you will gain experience with one of the most important techniques in modern molecular biology, the polymerase chain reaction (PCR), using the first commercially available PCR-based kit used in the identification of individuals in criminal and paternity cases. While learning these techniques, you will examine data, specific to a case study, generated using the AmpliType PM/DQA1 kit (Applied Biosystems). Following DNA isolation and amplification, the PM/DQA1 products from the known

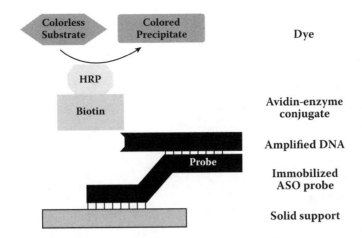

FIGURE 8

Detection of PCR Products Using the AmpliType PM/DQA1 Test and the Reverse Dot Blot Technology. *Note*: The nylon membrane strip contains the immobilized allele-specific oligonucleotide (ASO) probes. The amplified PCR products from the known and/or evidentiary samples are labeled with a streptavidin (avidin)–biotin–enzyme (horse radish peroxidase, or HRP) complex and allowed to hybridize to the immobilized probes. Following hybridization, a colorless substrate is added, and if the amplified DNA bound to the probe, the enzyme will convert the colorless substrate to a blue precipitate.

and evidentiary samples will be hybridized to the nylon strip containing the immobilized DNA probes, and the responses recorded to determine the various genotypes and profiles used for identification.

Equipment and Material

1. Ice in buckets
2. Chelex beads
3. 0.5, 1.0, and 1.5 ml Eppendorf or microcentrifuge tubes
4. Mineral oil (optional)
5. 15 ml polypropylene test tube
6. Double-distilled water
7. 30 or 50 ml conical tubes
8. Adjustable-volume digital micropipets (2–200 μl range)
9. Aerosol-resistant pipet tips
10. Perkin-Elmer Cetus GeneAmp Kit
11. Taq DNA polymerase (if not supplied in kit)
12. Disposable gloves
13. TAE buffer
14. TBE buffer
15. Agarose
16. Molecular weight markers (526 to 22,621 bp)
17. Ethidium bromide or coomassie blue

18. Bromophenol blue tracking dye

19. 125 ml Erlenmeyer flask

20. Microwave or hot plate

21. Incubator or water bath at 56°C

22. Electrophoresis systems (gel tray or combs)

23. Power pack or supply

24. Microcentrifuge

25. DNA thermal cycler

26. Polaroid camera with film

Procedure

The use of disposable gloves and aerosol-resistant pipet tips is highly recommended to prevent cross-contamination. There are several steps in the AmpliType PM/DQA1 test:

1. DNA is extracted from the known and evidentiary samples (see Exercise 1 for extraction procedures). Due to the increased level of sensitivity when using PCR, the number of cells required is significantly fewer than that required for RFLP analysis.

2. The DNA from the sample is amplified using PCR, tagged or labeled using the streptavidin (avidin)–biotin–enzyme complex described above, and allowed to hybridize to the PM/DQA1 probes bound to the nylon strips.
   Note: In RFLP, the target DNA is bound to the blot, and the probe DNA is added. For the PM/DQA1 test, the probe DNAs are immobilized on a nylon strip and the target DNA is added.

3. The pattern of blue dots determines the DNA type or profile at the HLA DQA1 locus. The pattern of blue dots in Figure 10 is an example of how the DQA1 profile is determined.

4. The AmpliType PM nylon strips contain 5 genetic loci: LDLR, GYPA, HBGG, D7S8, and GC. Each locus represents a distinct location or site on the DNA. When compared to the allelic variations for the DQA1 locus, the 5 PM loci have rather simple allelic variations. For example, there are only two alleles for the LDLR, GYPA, and D7S8 loci designated allele A and/or B. The HBGG and GC loci each have 3 possible alleles: A, B, and C. The "S" dot is the control dot. The response or signal at this site is designed to be less than that observed for the other dots. A faint response at the "S" dot indicates that a threshold has been met during the hybridization process rendering support for the other hybridization responses (Figure 11).

A Case Study

A man named Harry lived alone. Instead of cooking for himself, Harry usually went out for his meals. One evening, Harry decided to go to a restaurant for dinner. While he was out, his home was robbed. When Harry returned home from dinner, he saw a man driving away from his house in a van. Although Harry did not get a good look at the face of the burglar, he could identify him as a Caucasian male in his twenties. Harry called the police, and sometime before midnight they had determined that the stolen items included a flat-screen TV, a stereo CD player, a digital video camera, a desktop computer, and a laptop.

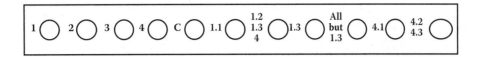

FIGURE 9

DQA1 Typing Strip. **Note:** Before hybridization, the nylon typing strips are colorless. The invisible dot identified as a "1" has a DNA probe for the allele variation designated number 1 for the DQA1 test. The invisible dot identified above the "2" contains the DNA probe for the "2" allele. The invisible dot above the 3 and 4 alleles contains the DNA probe for these alleles. The "C" dot is the control dot. The response or signal at this site is designed to be less than that observed for the other dots. A faint response at the "C" dot indicates that a threshold has been met during the hybridization process rendering support for the other responses. The invisible dot containing the 1 allele has several variations: the 1.1, 1.2, and 1.3 subtypes, also called alleles. The remaining invisible dots contain various combinations of allelic probes. This format has no specific or single dot or probe for the 1.2, 4.2, or 4.3 subtypes; thus, the design of the typing strip is unable to distinguish between these subtypes. The "circles" surrounding an area on the nylon strip are only present in these illustrations to indicate the location of the DNA probes.

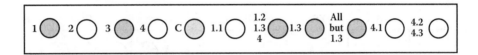

FIGURE 10

HLA DQA1 Typing Results. **Note:** Following hybridization, a colorless substrate is added, and if the amplified DNA bound to the probe the enzyme will convert the colorless substrate to a blue precipitate. The pattern of the "blue dots" is an example of a DQA1 typing strip that was hybridized with amplified DNA from a person with a DQA1 type of 1.3, 3. The typing results are interpreted from left to right. In this example, a faint response is observed at the "C," or control, dot. There is a response at the "1" allele; a response at the 3 allele (confirming a 3 allele); a response at the 1.2, 1.3, and 4 alleles; a response at the 1.3 allele (confirming a 1.3 allele); and a response at the "all but 1.3" alleles. Based on these responses, the DQA1 type or profile is 1.3, 3. The circles surrounding an area on the nylon strip are only present in these illustrations to indicate the location of the DNA probes.

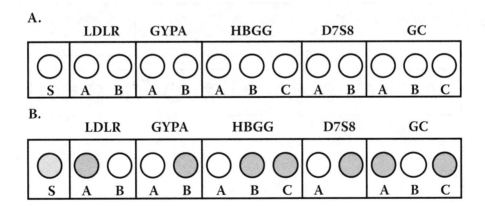

FIGURE 11

PM Typing Strip. **Note:** 1) Prior to hybridization, the nylon typing strips are colorless. The DNA probes are immobilized onto the nylon strip and identified with the corresponding locus (e.g., LDLR and GYPA). The circles surrounding an area on the nylon strip are only present in these illustrations to indicate the location of the DNA probes. Each locus will have 2 or 3 invisible dots that represent the allelic variations at that locus. 2) In this example, a faint response is observed at the "S," or control, dot. There is a response at the "A" allele of the LDLR locus, a response at the B allele of the GYPA locus, two responses at the B and C alleles of the HBGG locus, a response at the B allele of the D7S8 locus, and a response at the A and C alleles of the GC locus. Based on these responses, the AmpliType PM type or profile is AA/BB/BC/BB/AC or A/B/BC/B/AC.

During the investigation, the police located the digital video camera and the laptop at a local pawnshop. The pawnshop owner reviewed his sales records and provided the investigators on the case with several names whom he thought sold him the stolen merchandise. Sometime later and after careful questioning, the investigators had identified three male suspects who often sold "stolen" merchandise to local area pawn-shops. Two Caucasian males in their late twenties were "picked up" for questioning. Both male suspects had their blood drawn and sent to the local forensic laboratory for DNA analysis. The digital video camera and the laptop, the only two stolen items recovered, were also sent to the laboratory for analysis. The forensic DNA analyst subjected the swab of the digital video camera and a small drop of blood, found on the laptop, to AmpliType PM/DQA1 analysis. Reference samples from each of the two suspects were also subjected to the PM/DQA1 analysis. The results are shown below.

Results

Suspect #1

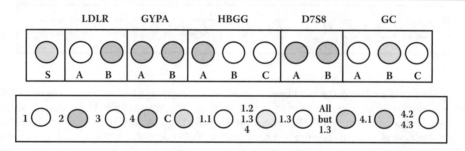

Suspect #2

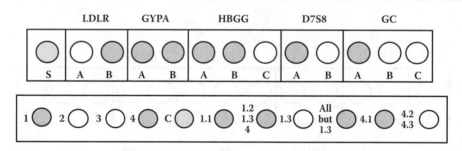

Swab from Digital Video Camera

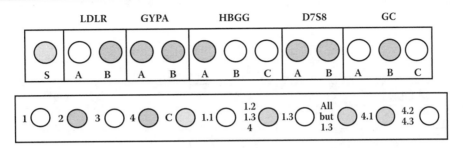

Blood from Laptop

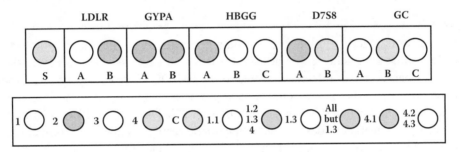

Interpreting Your AmpliType PM/DQA1 Typing Results

Basis of the AmpliType PM/DQA1 Typing System

- Six loci were amplified simultaneously with a pair of primers for each locus.

- At each locus, 2 or more alleles have been observed in a given population.

- An individual is either homozygous or heterozygous for any of the two possible alleles:
  - LDLR (low-density lipoprotein receptor): alleles A and/or B
  - GYPA (glycophorin A): alleles A and/or B
  - HBGG (hemoglobin G gamma globulin): alleles A, B, and/or C
  - D7S8 (chromosome 7): alleles A and/or B
  - GC (group-specific component): alleles A, B, and/or C
  - DQA1 (human leukocyte antigen): alleles 1.1, 1.2, 1.3, 2, 3, 4.1, 4.2, and/or 4.3

- A control spot (an "S" spot on the PM strip and a "C" spot on the DQA1 strip) is included on each of the nylon test strips. The control spots are included to determine the minimum amount of response needed (i.e., the amount signal from the enzymatic reaction) for a positive reaction. Any signal or spot lighter than the control spot is considered a negative response, and the test invalid. Any signal or spot darker than the control spot is considered a positive response and valid.

Record your observations or the PM/DQA1 DNA typing profiles of the known reference samples (i.e., Suspects #1 and #2) and the evidentiary samples (i.e., swab from camera and blood from the laptop) using the table below.

AmpliType PM/DQA1 Typing Results

Sample	Polymarker (PM) Loci					DQA1
	LDLR	GYPA	HBGG	D7S8	GC	
Suspect #1						
Suspect #2						
Swab from Camera						
Blood from Laptop						

Your group will work together to interpret the data generated from the AmpliType PM/DQA1 typing analysis. First, determine the allelic profile at each locus for all samples. Then compare the DNA typing results from each suspect's known reference sample to the PM/DQA1 profiles generated from the swab of the digital video camera and the blood from the laptop. For each sample, determine if the DNA pattern appears to be heterozygous or homozygous at the different PM/DQA1 loci. If an allelic response at any locus from the evidentiary samples does not match the allelic responses observed from either of the known reference samples, then that suspect(s) is excluded. If all of the alleles for each locus "match" (i.e., between the known reference sample and the evidentiary sample), then that individual (Suspect #1 or #2) cannot be excluded as being the contributor to the questioned samples. When considering the results, assume that all controls (e.g., positive and negative) responded appropriately. After completing your analysis, answer the questions below.

Questions

1. What is the genotype (at all 6 loci) of Suspect #1? Explain your reasoning.

2. What is the genotype (at all 6 loci) of Suspect #2? Explain your reasoning.

3. Is Suspect #1 included or excluded as the contributor of the swab from the digital video camera? From the blood from the laptop? Explain your reasoning.

4. Is Suspect #2 included or excluded as the contributor of the swab from the digital video camera? From the blood from the laptop? Explain your reasoning.

5. Do the DNA profiles from the swab of the camera and the blood from the laptop match? Explain your answer.

6. If, in analyzing evidence from a crime scene using the AmpliType PM/DQA1 system, two or more suspects showed the same allelic pattern as that seen in this case study, what could you do to resolve the question?

10

Exercise 7

Polymerase Chain Reaction (PCR)–Based Tests: The D1S80 System

Introduction

The nucleus of a human somatic cell contains 6.4 billion base pairs of DNA. Specific portions of this DNA encode for over 100,000 genes, whereas the remainder of the DNA is comprised of noncoding regions often referred to as "junk DNA." Approximately 99.5% of the DNA code is identical for all people. The remaining 0.5% is of interest to forensic scientists because of the variations in the DNA that exist between individuals. Consequently, forensic DNA analysis focuses on these differences or hypervariable regions of DNA between individuals. One such form of variability is called a variable number of tandem repeats (or VNTR). The most forensically characterized VNTR is the polymorphism at the locus denoted D1S80 (Nakamura et al., 1988). This locus, located on human chromosome 1, is composed of repeating units of DNA segments that are 16 nucleotides in length (Kasai et al., 1990). The number of tandem repeats varies from one individual to the next, and is known to range from 15 to over 41. This variability in the number of tandem repeats is the basis for identification in the D1S80 typing system.

The D1S80 system was used by some forensic laboratories to complement the existing PCR-based assays, the AmpliType PM/DQA1 system, and, in many instances, RFLP analysis. Thus, the advent of PCR offered a viable alternative to RFLP analysis, especially if the questioned or evidentiary sample contained an insufficient quantity of DNA or the DNA was suspected to be degraded. The D1S80 system was amenable to PCR amplification due to the relatively small size of the locus with amplified products less than 700 bp in length. Moreover, the D1S80 system combined the advantages inherent with any PCR-based system with the greater variations seen in RFLP analysis. The use of PCR reduced assay time and cost, and served as an alternative or as a complement to other DNA testing. Similar to the DQA1 test, only one locus is analyzed in this system, thus limiting the D1S80's power of discrimination. However, when used in conjunction with other DNA tests, the power of discrimination increased dramatically.

Objective

In this exercise, you will gain experience with one of the most important techniques in modern molecular biology, the polymerase chain reaction (PCR), while using the D1S80 typing system. While learning these techniques, you will extract DNA from cells in your mouth (buccal swabs) and/or from a human cell line, incubate the isolated DNA with appropriate PCR reagents, and amplify the alleles at the D1S80 locus. Following amplification, the amplified products will be separated and identified using agarose gel electrophoresis.

Equipment and Material

1. 0.5, 1.0, and 1.5 ml Eppendorf or microcentrifuge tubes
2. TAE buffer
3. Chelex beads
4. Disposable gloves
5. 15 ml polypropylene test tube
6. Double-distilled water
7. 30 or 50 ml conical tube
8. Mineral oil (optional)
9. Aerosol-resistant pipet tips
10. Perkin-Elmer Cetus GeneAmp kit
11. Taq DNA polymerase (if not supplied in kit)
12. Adjustable-volume digital micropipets (2–200 µl range)
13. Ice in buckets
14. Primers (D1S80) 0.1 µg/µl
    A. 5′ GAA ACT GGC CTC CAA ACA CTG CCC 3′
    B. 5′ GTC TTG GAG ATG CAC GTG CCC CTT GC 3′
15. Genomic DNA from human cell lines* (10 ng/µl)
    A. HEP G2: hepatocellular carcinoma (liver)
    B. HTB 180 NCI-H345: small cell carcinoma, lung
    C. CCL 86: Raji Burkitt lymphoma
    D. HTB 184 NCI-H510A: small cell carcinoma, extrapulmonary origin
    E. CRL 1905 H: normal skin cell line
    F. HeLa: epithelial carcinoma cell line
    G. K562: erythromyeloblastoid leukemia cell line, chronic myeloid leukemia
16. Agarose
17. TBE buffer
18. Bromophenol blue tracking dye

---

* Examples of human cell lines that can be used to demonstrate a D1S80 profile.

19. 125 ml Erlenmeyer flask

20. Ethidium bromide or coomassie blue

21. Molecular weight markers (526 to 22,621 bp)

22. Incubator or water bath at 56°C and 100°C

23. Electrophoresis systems (gel tray or combs)

24. Power pack or supply

25. DNA thermal cycler

26. Microwave or hot plate

27. Microcentrifuge

28. Tabletop clinical centrifuge

29. Polaroid camera with film

Procedure

The use of disposable gloves and aerosol-resistant pipet tips is highly recommended to prevent cross-contamination. Helpful organizational sheets are provided at the end of the exercise.

Collection of Cells (e.g., buccal cells, liquid blood, and cultured human cells)

1. Label a 15 ml polypropylene test tube and the top of a 1.5 ml Eppendorf or microcentrifuge tube with your name and any other appropriate information.

2. Pipet 10 ml of cultured human cells (maximum 5×10^6 cells/ml) or liquid sample into the polypropylene test tube. The cultured human cells can serve as the positive control or as the "mock" evidentiary sample. For buccal cells, pour 10 ml of saline solution into your mouth, and vigorously swish against your cheeks for 10 sec. Expel saline solution back into the labeled 15 ml polypropylene test tube over the sink.

Or,

if sterile swabs are available, place the swab inside your mouth, and press it firmly against the inside of your cheek. Roll the swab back and forth over the inside surface of your cheek at least 10 times. Repeat on the other cheek. Place the swab into a labeled 15 ml test tube containing 10 ml saline solution.

Concentrate Cells by Centrifugation

3. Centrifuge the samples at $300 \times g$ for 5 min. The cells will form a firm pellet at the bottom of the tube.

4. Decant the saline supernatant into a liquid waste container. Be careful not to disturb the cell pellet at the bottom of the tube.

5. Add 500 μl of Chelex beads into the 15 ml test tube containing the cell pellets. Resuspend cells with Chelex either by pipetting in and out several times or by tapping with your finger.

6. Transfer a 500 μl aliquot of the cell–Chelex slurry into a sterile 1.5 ml Eppendorf tube. Make sure the Eppendorf tube is labeled.

Lysing the Cells and Collecting the DNA

7. Place the capped (closed) Eppendorf tubes in a "float," and place in a boiling water bath for 10 min.

8. After the heat treatment, place the samples on ice for 5 min.

9. Place the Eppendorf tubes containing the lysed cells in a microcentrifuge, and spin at the maximum speed for 1 min. The pellet contains the Chelex beads bound to the denatured proteins. The supernatant contains the DNA.

10. Using a sterile pipet tip, transfer all of the clear supernatant to a fresh 1.5 ml Eppendorf tube. Label the tube, and place on ice until you are ready to proceed to the next step.

Setting Up the PCR Amplification

11. Label 3 0.5 ml microcentrifuge tubes as follows: 1) "DNA/human cell line—positive control"; 2) "buccal swab"; and 3) "no DNA—negative control."

12. Dispense 10 μl of appropriate genomic DNA (10 ng/μl) into the labeled tubes and 10 μl of distilled H_2O into the tube labeled "no DNA."

13. Prepare the following "master mix" for 6.5 reactions (see Table 3). To determine the final volume of the master mix needed for all of the reactions, calculate the required amount of each component of the PCR master mix.

14. Add 15 μl of the master mix to the 3 tubes containing genomic DNA, mix, and overlay with 50 μl of mineral oil if needed. Depending on the thermal cycler model used, this step may be omitted.

15. Place the reaction tubes into a thermal cycler programmed to run at least 35 cycles with the following parameters:

Step 1	2.0 min	94°C	Denaturation
Step 2	1.0 min	94°C	Denaturation
Step 3	1.0 min	64°C	Annealing
Step 4	1.5 min	70°C	Extension
Step 5	Repeat Steps 2–4 for a total of 35 times.		
Step 6	Soak	4°C	

TABLE 3
PCR Master Mix for the D1S80 Locus

PCR Master Mix Component	Volume	Final Concentration
10X PCR buffer	16.25 μl	1X
10 mM dNTP	13.00 μl	800 μM
D1S80 primer (a)	3.25 μl	50 ng/reaction
D1S80 primer (b)	3.25 μl	50 ng/reaction
Taq DNA polymerase	3.25 μl	
Distilled H_2O	58.50 μl	
Total volume	97.50 μl	

Note: The parameters outlined above may vary according to the thermal cycler used for the PCR amplification.

16. Start or "run" the PCR incubation reaction. When the thermal cycler's program is completed (approximately 3 hr), the tubes containing the PCR products will be removed by your instructor and stored at 0°C.

Visualization of the D1S80 PCR Products

17. Because small PCR products (or DNA fragments) in the range of 200 to 600 base pairs are expected, a high concentration of agarose is required for adequate separation during gel electrophoresis. Prepare the agarose gel as described in Step 2 of the "Procedure" section of Exercise 4; however, TBE buffer is preferred (instead of TAE buffer) for separation of DNA molecules less than 1 kb in size.

18. Remove 16 µl of the PCR product(s), and place into separate microcentrifuge tubes. Add 4 µl of loading or tracking dye to each tube.

19. The DNA sample, mixed with loading buffer (6 µl total volume), is pipetted into the well with the gel submerged in TBE buffer. Be careful not to push the pipet tip through the bottom of the well in the gel.

20. Include on your gel the K562 DNA standards (concentration determined previously). Load the gel as follows:

Lane 1	K562 DNA (positive control)	16 µl
Lane 2	Molecular weight markers	16 µl
Lane 3–18	DNA from buccal swabs	16 µl
Lane 19	Molecular weight markers	16 µl
Lane 20	No DNA control	16 µl

21. Set the voltage (100 volts), and turn on the power supply. "RUN" the samples until the bromophenol blue tracking dye has moved 1–2 cm from the origin (i.e., well) or until the dye front is approximately 2 cm from the bottom of the gel. This should take less than 1 hr.

22. Stain the gel in SYBR green or ethidium bromide (if not added previously to the agarose gel or reservoir buffer). Photograph the gel, and determine the allele response and size at the D1S80 locus.

Sample Setup for Thermal Cycler

Analyst: _____ Date: _____ Gel No: _____

Thermal Cycler: _____   Start Time: _____

Date of Last Calibration: _____   System: _____

A1	A2	A3	A4	A5	A6	A7	A8	A9	A10	A11	A12
B1	B2	B3	B4	B5	B6	B7	B8	B9	B10	B11	B12
C1	C2	C3	C4	C5	C6	C7	C8	C9	C10	C11	C12
D1	D2	D3	D4	D5	D6	D7	D8	D9	D10	D11	D12
E1	E2	E3	E4	E5	E6	E7	E8	E9	E10	E11	E12
F1	F2	F3	F4	F5	F6	F7	F8	F9	F10	F11	F12
G1	G2	G3	G4	G5	G6	G7	G8	G9	G10	G11	G12
H1	H2	H3	H4	H5	H6	H7	H8	H9	H10	H11	H12

Results

D1S80 GEL

Analyst: _____   Lab Number: _____

Date: _____

Well No.	Sample	DNA (ng)
1	K562 DNA	
2	Molecular Weight Marker	
3	DNA from Buccal Swab	
4	"	
5	"	
6	"	
7	"	
8	"	
9	"	
10	"	
11	"	
12	"	
13	"	
14	"	
15	"	
16	"	
17	"	
18	"	
19	Molecular Weight Marker	
20	No DNA Control	

Reagents	Lot Number	Source
Agarose		
1X TBE (Gel Buffer)		
1X TBE (Tank Buffer)		
Loading Buffer		
Ethidium Bromide		
Visual Marker		

Gel Electrophoresis

Time on:	Voltage:	mAMPs:
Time off:	Voltage	mAMPs:

Gel Prepared By: _____ Date: _____

Reporting Form

Tape Your Gel Photo Here

Lane #1 _____

Lane #2 _____

Lane #3 _____

Lane #4 _____

Lane #5 _____

Lane #6 _____

Lane #7 _____

Lane #8 _____

Lane #9 _____

Lane #10 _____

Lane #11 _____

Lane #12 _____

Lane #13 _____

Lane #14 _____

Lane #15 _____

Lane #16 _____

Lane #17 _____

Lane #18 _____

Lane #19 _____

Lane #20 _____

D1S80 GEL

Analyst: _____   Lab Number: _____

Date: _____

Well No.	Sample	DNA (ng)
1		
2		
3		
4		
5		
6		
7		
8		
9		
10		
11		
12		
13		
14		
15		
16		
17		
18		
19		
20		

Interpreting Your Results

Your group will work together to interpret the photograph of your gel with the test results. Attach the gel photograph to the Reporting Form (see previous). For each lane, state the approximate size of each fragment and if the person appears to be heterozygous or homozygous at the D1S80 locus. Use the table below to record your observations. Also, examine the other groups' test results, and attempt to determine the number of responses or the different alleles present in the class. Are any of the profiles similar? Are any two genotypes the same?

Basis of the D1S80 Typing System

- The locus consists of repeating units of segments of DNA that are 16 nucleotides in length.
- The number of tandem repeats varies from one individual to the next.
- The alleles range from 15 to over 41. This variability in the number of tandem repeats is the basis for identification in the D1S80 system.

D1S80 Typing Results

Sample	D1S80 Locus
1	
2	
3	
4	
5	
6	
7	
8	
9	
10	
11	
12	
13	
14	
15	
16	
17	
18	
19	
20	

Questions

1. What are the 3 steps in the PCR process?

2. What are the necessary components of the PCR process?

3. How are the PCR products analyzed to ensure that the PCR reaction was successful?

4.   What is the "power of discrimination" with the D1S80 typing system?

Chapter 11

Experiment 8
Polymerase Chain Reaction (PCR)–Based Tests: Short Tandem Repeat (STR) Analysis

Introduction

A DNA segment that appears more than once on the same chromosome is known as a repeat. Human genomes contain 5–10% of such repetitive sequences that occur in tandem or adjacent to each other. These repetitive sequences vary in size and length and show sufficient variability among individuals in a population. Regions of DNA that contain these short repeated segments are referred to as short tandem repeats (STRs) and are important markers for human identity testing in the forensic community.

There are literally thousands of STR markers scattered throughout the human genome, and they occur, on average, in one in every 10,000 nucleotides. The DNA sequence repeated in an STR motif is usually from 2 to 7 base pairs (bp), with 4 bases being the preferred size for forensic systems (Edwards et al., 1991, 1992; Warne et al., 1991). An example of a 4 bp or a tetranucleotide repeat is shown below, where the TCTA motif is repeated 4 times.

…ATGTGA TCTA TCTATCTATCTATTGG…

PCR-based STR systems offer many advantages over earlier DNA typing techniques (e.g., restriction fragment length polymorphisms, or RFLP). STR systems provide a rapid and sensitive method to evaluate small amounts (1 ng) of human DNA. This small amount of DNA needed for STR systems is 50 times less than what is normally required for RFLP analysis. Also, the repeating sequences in an STR are relatively short, with the entire STR strand or allele generally less than 400 bp in length. This short length renders STR systems amendable to the analysis of samples suspected of being degraded. STR analysis often allows the DNA analyst to recover a complete DNA profile even from an evidentiary sample that was exposed to unfavorable conditions (e.g., body or stains subject to extreme decomposition). This is in sharp contrast to

RFLP systems, which require a large sample size for analysis and full-length fragments, which often consist of thousands of bases, to generate a complete DNA profile.

STRs and corresponding loci are easily amplified by PCR. Further, PCR amplification of many different STR loci is commonly performed simultaneously in the same tube. The simultaneous amplification of 2 or more loci is commonly known as multiplexing, or multiplex PCR. For a multiplexing reaction to be successful, the system must be designed to ensure that the sizes of the amplified products do not overlap, thereby allowing each STR allele for a specific locus to be clearly visualized on a gel or by capillary electrophoresis. This "requirement design" of overlapping fragments became less important with the development of multiple color detection systems.

Different detection methods are available to visualize the STR products. The STR loci and corresponding alleles may be separated by gel electrophoresis and detected using ethidium bromide and silver staining or exotic dyes (e.g., SYBR green). Several STR systems have been developed where fluorescent dyes or labels are used to detect the STR alleles either during separation (i.e., capillary electrophoresis) or after separation (i.e., gel electrophoresis). The resulting STR profiles are routinely interpreted by direct comparison to DNA standards, allelic ladders (an artificial mixture of common alleles present in the human population for a particular STR marker or locus), and reference standards (known DNA profiles from the victim and suspect). Probability calculations are determined based upon classical population genetic principles.

For STR markers to be effective across various jurisdictions, a common set of standardized markers is used. Currently, the forensic scientific community in the United States has established a set of 13 core STR loci that, in turn, can be entered into a national database known as the Combined DNA Index System (CODIS), a collection of DNA profiles from known offenders. A summary of the 13 CODIS loci is contained in Table 4.

TABLE 4
Information on the 13 Core Short Tandem
Repeat Loci Listed in CODIS

STR Locus	Chromosome Number	Sequence
FGA	4	CTTT
vWA	12	[TCTG][TCTA]
D3S1358	3	[TCTG][TCTA]
D21S11	21	[TCTA][TCTG]
D8S1179	8	TATC
D7S820	7	GATA
D13S317	13	TATC
D5S818	5	AGAT
D16S539	16	GATA
CSF1PO	5	AGAT
TPOX	2	AATG
THO1	11	TCAT
D18S51	18	AGAA

Objective

In this exercise, you will extract DNA from cells in your mouth (buccal swabs) and/or from a human cell line, incubate the isolated DNA with appropriate PCR reagents, and amplify the alleles at multiple STR loci using the GenePrint STR Systems (Promega Corporation, Madison, Wisconsin). Following amplification, the amplified STR products will be separated and identified using agarose gel electrophoresis.

Equipment and Material

1. 0.5, 1.0, and 1.5 ml Eppendorf or microcentrifuge tubes
2. 30 or 50 ml conical tubes
3. Mineral oil (optional)
4. 15 ml polypropylene test tube
5. Double-distilled water
6. Sterile cotton swabs
7. Adjustable-volume digital micropipets (2–200 μl range)
8. Aerosol-resistant pipet tips
9. GenePrint STR Systems CSF1PO, TPOX, and TH01 (Promega Corporation)
10. Taq DNA polymerase (not supplied in kit)
11. Disposable gloves
12. Ice in buckets
13. Genomic DNA from human cell lines* (10 ng/μl)
    A. HEP G2: Hepatocellular carcinoma (liver), male
    B. HTB 180 NCI-H345: Small cell carcinoma, lung
    C. CCL 86: Raji Burkitt lymphoma
    D. HTB 184 NCI-H510A: Small cell carcinoma, extrapulmonary origin
    E. CRL 1905 H: Normal skin cell line
    F. HeLa: Epithelial carcinoma cell line
    G. K562: Erythromyeloblastoid leukemia cell line, chronic myeloid leukemia
14. Molecular weight markers (526 to 22,621 bp)
15. Ethidium bromide or coomassie blue
16. Bromophenol blue tracking dye
17. 125 ml Erlenmeyer flask
18. TBE buffer
19. TAE buffer
20. Agarose
21. DNA thermal cycler

---

* Examples of human cell lines that can be used to demonstrate an STR profile and serve as a positive control.

22.   Electrophoresis systems (gel tray or combs)

23.   Power pack or supply

24.   Microwave or hot plate

25.   Incubator or water bath at 56°C

26.   Microcentrifuge

27.   Tabletop clinical centrifuge

28.   Polaroid camera with film

Procedure

To prevent cross-contamination, the use of disposable gloves and aerosol-resistant pipet tips is highly recommended. Helpful organizational sheets are provided at the end of the exercise.

1.   Refer to Exercise 1 for the methods and/or steps used in the collection and concentration of cells and for methods outlining cell lysis and the collection of DNA.

Setting Up the PCR Amplification

1.   Thaw the PCR reagents (STR 10X buffer and STR 10X primer pairs), and keep on ice. These reagents will be combined to form the PCR "master mix" for the multiplex reactions.

2.   Determine the number of reactions to be set up. Positive and negative controls should also be included when determining the number of reactions.

3.   For each reaction, label one sterile 0.5 ml microcentrifuge tube, and place into a rack.

4.   To determine the final volume of the master mix needed for all of the reactions, calculate the required amount of each component of the PCR master mix (see Table 5). Multiply the volume (μl) per sample by the total number of reactions (from Step 2) to obtain the final volume (μl). To compensate for pipetting error, add enough components to the master mix for two additional reactions.

5.   In the order listed in Table 5, add the final volume of each component to a sterile microcentrifuge tube. Once completed, mix the components gently, and place on ice.

6.   Add 22.50 μl of the PCR master mix to each tube (from Step 4, above), and place on ice.

7.   For amplification, add the appropriate volume (use 5 ng) of template DNA (extracted DNA from the buccal swabs and the human cell lines) to each reaction tube.

8.   For the positive control, pipet 2.5 μl of genomic DNA (5 ng of human cell line DNA) into a 0.5 ml microcentrifuge tube containing 22.5 μl of the PCR master mix.

9.   For the negative control, pipet 2.5 μl of sterile water (not template DNA) into a 0.5 ml microcentrifuge tube containing 22.5 μl of the PCR master mix.

10.   Add 1 drop of mineral oil to each microcentrifuge tube to prevent evaporation. Close the tubes, and centrifuge briefly (5 sec). Depending on the thermal cycler model used, this step may be omitted.

11.   Place the reaction tubes into a thermal cycler programmed to run at least 35 cycles with the following parameters:

TABLE 5
Multiplex Reactions Containing Three Loci

PCR Master Mix Component	Volume per Sample (μl)	Number of Reactions	Final Volume (μl)
Sterile water	17.35		
STR 10X buffer	2.50		
Multiplex 10X primer pair mix	2.50		
Taq DNA polymerase (at 5u/µl)*	0.15 (0.75u)		
Total volume	**22.50**		

Note: If the DNA is stored in TE buffer, the volume of the DNA sample should not exceed 20% of the final volume because components of the buffer compromise PCR amplification efficiency and quality. This rule does not apply to DNA stored in sterile water.

* The volumes or values given for Taq DNA polymerase assume a concentration of 5u/Fl. If the final volume is less than 0.5 µl, the enzyme can be diluted with STR 1X buffer, and then a larger volume added. Because the enzyme cannot be stored diluted, only prepare the amount that you will need. The amount of sterile water can be adjusted accordingly so that the final volume of each reaction is 25 Fl.

Step 1	2.0 min	94°C	Denaturation
Step 2	1.0 min	94°C	Denaturation
Step 3	1.0 min	64°C	Annealing
Step 4	1.5 min	70°C	Extension
Step 5	Repeat Steps 2–4 for a total of 35 times.		
Step 6	Soak	4°C	

Note: The parameters outlined above may vary according to the thermal cycler used for the PCR amplification.

12. Start or "run" the PCR incubation reaction. When the thermal cycler's program is completed (approximately 3 hr), the tubes containing the PCR products will be removed by your instructor and stored at 0°C.

Visualization of the STR PCR Products

1. Because small PCR products (or DNA fragments) in the range of 150 to 400 base pairs are expected, a high concentration of agarose is required for adequate separation during gel electrophoresis. Prepare the agarose gel as described in Step 2 of the "Procedure" section of Exercise 4.

2. Remove 2.5 µl of the PCR product(s), and place into separate microcentrifuge tubes. Add 2.5 µl of STR 2X Loading Solution (supplied with kit) to each tube.

3. Add 2.5 µl (50 ng) of pGEM DNA markers (supplied with kit) to 2.5 µl of STR 2X loading solution.

Note: The pGEM DNA markers are visual standards used to confirm allelic size ranges for each locus. The markers consist of 15 DNA fragments with weights (bp) of 2645, 1605, 1198, 676, 517, 460, 396, 350, 222, 179, 126, 75, 65, 51, and 36.

4. Add 2.5 μl of the STR allelic ladder (supplied in kit) to 2.5 μl of STR 2X loading solution for each ladder lane (at least 2 per gel).

5. Using a different pipet tip for each sample, load the DNA samples, mixed with loading solution (5 μl total volume), into the wells with the gel submerged. Be careful not to push the pipet tip through the bottom of the well in the gel.

6. Include on your gel the positive and negative controls. Load the gel as follows:

Lane 1	pGEM markers	5 μl
Lane 2	STR allelic ladder*	5 μl
Lane 3	Human cell line DNA (positive control)	16 μl
Lane 4	Negative control (no DNA)	16 μl
Lanes 5–11	DNA from buccal swabs	16 μl
Lane 12	STR allelic ladder	5 μl
Lanes 13–19	DNA from buccal swabs	16 μl
Lane 20	STR allelic ladder	5 μl

    * For ease of interpretation, the allelic ladders can be run in lanes adjacent to each sample.

7. Set the voltage (100 volts), and "RUN" the samples until the bromophenol blue tracking dye has moved 1–2 cm from the origin (i.e., well) or until the dye front is approximately 2 cm from the bottom of the gel. This should take less than 1 hr.

8. Photograph the gel, and determine the allele response and size at each locus. Direct comparison between the allelic ladders and amplified samples of the same locus should allow for the numerical assignment of each allele (see Table 6 and Figure 12).
    Artifacts or unidentifiable DNA bands may be detected with this system. For questions related to these STR "by-products," refer to the "Troubleshooting Guide" in the technical manual for the Geneprint System.

TABLE 6
Locus-Specific Information for the Geneprint STR Systems
(CSF1PO, TPOX, and TH01)

Component Loci	Allelic Ladder Size Range (Bases)	STR Ladder Alleles (Number of Repeats)	Other Known Alleles	K562 DNA Allele Sizes
CSF1PO	295–327	7, 8, 9, 10, 11, 12, 13, 14, 15	6	9, 10
TPOX	224–252	6, 7, 8, 9, 10, 11, 12, 13	None	8, 9
THO1	179–203	5, 6, 7, 8, 9, 10, 11	9.3	9.3, 9.3

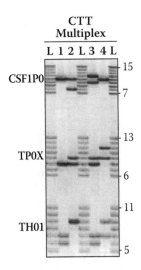

FIGURE 12

Profiles from the GenePrint STR Systems (CSF1PO, TPOX, and TH01). *Note*: Genomic DNA (Lanes 1–4) was amplified using the CTT Multiplex STR System, separated in a 4% polyacrylamide denaturing gel, and detected using silver stain. The lanes labeled "L" contain the allelic ladders for each locus (i.e., CSF1PO, TPOX, and TH01). The numbers to the right of the image indicate the smallest to the largest number of repeats or numerical designation for each allele. *Source*: Courtesy of the Promega Corporation.

Forensic DNA Analysis: A Laboratory Manual

Sample Setup for Thermal Cycler

Analyst: _____ Date: _____ Gel No: _____

Thermal Cycler: _____ Start Time: _____

Date of Last Calibration: _____ System: _____

A1	A2	A3	A4	A5	A6	A7	A8	A9	A10	A11	A12
B1	B2	B3	B4	B5	B6	B7	B8	B9	B10	B11	B12
C1	C2	C3	C4	C5	C6	C7	C8	C9	C10	C11	C12
D1	D2	D3	D4	D5	D6	D7	D8	D9	D10	D11	D12
E1	E2	E3	E4	E5	E6	E7	E8	E9	E10	E11	E12
F1	F2	F3	F4	F5	F6	F7	F8	F9	F10	F11	F12
G1	G2	G3	G4	G5	G6	G7	G8	G9	G10	G11	G12
H1	H2	H3	H4	H5	H6	H7	H8	H9	H10	H11	H12

Results

STR Test Gel

Analyst: _____   Lab Number: _____

Date: _____

Well No.	Sample	DNA (ng)
1	pGEM Marker	
2	STR Allelic Ladder	
3	Human Cell Line (Positive Control)	
4	No DNA (Negative Control)	
5	DNA from Buccal Swabs	
6	"	
7	"	
8	"	
9	"	
10	"	
11	"	
12	STR Allelic Ladder	
13	"	
14	"	
15	"	
16	"	
17	"	
18	"	
19	"	
20	STR Allelic Ladder	

Reagents	Lot Number	Source
Agarose		
1X TBE (Gel Buffer)		
1X TBE (Tank Buffer)		
Loading Buffer		
Ethidium Bromide		
Visual Marker		

Gel Electrophoresis

Time on:	Voltage:	mAMPs:
Time off:	Voltage	mAMPs:

Gel Prepared By: _____   Date: _____

Reporting Form

Tape Your Gel Photo Here

STR Test Gel

Analyst: _____   Lab Number: _____

Date: _____

Well No.	Sample	DNA (ng)
1		
2		
3		
4		
5		
6		
7		
8		
9		
10		
11		
12		
13		
14		
15		
16		
17		
18		
19		
20		

Interpreting Your STR Results

Your group will work together to interpret the photograph of your gel with the test results. Attach the gel photograph to the reporting form (see previous page). Determine the allele response and size at each locus. Direct comparison between the allelic ladders and amplified samples of the same locus should allow for the numerical assignment of each allele (see Figure 12 and Table 6).

For each lane, state the approximate size of each fragment and if the person appears to be heterozygous or homozygous at the different STR loci. Also, examine the other groups' test results, and attempt to determine the number of responses or the different alleles present in the class. Are any of the profiles similar? Are any two genotypes the same?

STR Typing Results

Sample	LOCUS		
	CSF1PO	TPOX	THO1
1			
2			
3			
4			
5			
6			
7			
8			
9			
10			
11			
12			
13			
14			
15			
16			
17			
18			
19			
20			

Questions

1. What are the advantages of using STR systems versus some of the earlier DNA typing techniques?

2. STR loci chosen for use in the forensic community have many characteristics. Describe 3 favorable characteristics of STR loci.

3. What are some of the challenges that a forensic DNA analyst confronts with STR typing?

4.   Why are STRs preferred genetic markers?

12

Experiment 9

Polymerase Chain Reaction (PCR)–Based Tests: Short Tandem Repeat (STR) Analysis to Determine Paternity (a Case Study)

Introduction

DNA typing is the most accurate form of paternity testing possible. DNA typing can indicate with 100% certainty if the tested male is excluded as the biological father or will demonstrate with a high degree of scientific certainty (i.e., greater than 99.9% probability) if the tested male is the biological father. DNA paternity tests can be used to answer questions and/or issues such as the following:

- Paternity or maternity identification and verification
- Child support and custody disputes
- Suspected incest cases
- Inconclusive paternity results from other methods
- Single-parent cases where paternity or maternity is in question
- Newborn testing
- Prenatal paternity cases
- Identification of father in surrogate mother cases
- Estate or trust disputes

Parentage testing is performed by collecting biological samples (e.g., blood or buccal swabs) from the mother, the child, and the alleged biological father. For newborns, testing can be performed using umbilical blood from the umbilical cord. In unusual circumstances, DNA can be collected from other sources, as previously described (see "Types of Biological Samples"). DNA testing is based on genetic information that is passed on from the parents to their children (see Table 7). In cases where the alleged father is unavailable for testing, partial pedigree analysis can be conducted using DNA samples from the parents of the alleged father. If necessary, siblings of the alleged father can also be used.

Objective

In this exercise, DNA has been extracted from samples (i.e., buccal swabs) collected from the mother, the child, and the alleged father. The isolated DNA was amplified at multiple STR loci using PCR and the GenePrint STR Systems (Promega Corporation). Following amplification, the amplified STR products were separated and identified using polyacrylamide gel electrophoresis (PAGE). In this exercise, you will be analyzing data, specific to a case study, generated using PAGE. As the amplified STR products are separated, the fluorescently labeled DNA molecules are excited by the laser light source, and the emission is captured by a detection system and recorded as a chemilumigraph (e.g., bar code format) by the computer.

Equipment and Material

1. 0.5, 1.0, and 1.5 ml Eppendorf or microcentrifuge tubes

2. Agarose

3. Mineral oil (optional)

4. 15 ml polypropylene test tube

5. Double-distilled water

6. Ice in buckets

7. Adjustable-volume digital micropipets (2–200 µl range)

8. Aerosol-resistant pipet tips

9. GenePrint STR Systems (PowerPlex 16 BIO System) (Promega Corporation, Madison, Wisconsin)

10. Taq DNA polymerase (not supplied in kit)

11. Disposable gloves

12. TAE buffer

13. Genomic DNA (10 ng/µl) from mother, child, and alleged father

14. Molecular weight markers (526 to 22,621 bp)

15. Ethidium bromide or coomassie blue

16. Bromophenol blue tracking dye

17. 125 ml Erlenmeyer flask

18. TBE buffer (loading buffer)

19. 30 or 50 ml conical tubes

20. Incubator or water bath at 56°C

21. Electrophoresis systems (gel tray or combs)

22. Power pack or supply

TABLE 7
PowerPlex 16 BIO Typing Results
DNA was extracted from samples collected from the mother, the child, the boyfriend, and the alleged father and amplified at the STR loci using the PowerPlex 16 BIO System.

Locus	15-Year-Old Mother	Child	18-Year-Old Boyfriend	Alleged Father
FGA	21, 23	22, 23	21, 22	20, 22
TPOX	8, 10	8, 11	11, 11	11, 11
D8S1179	12, 13	12, 13	13, 13	13, 13
vWA	14, 18	14, 15	17, 18	14, 15
Penta E	10, 11	11, 16	10, 11	14, 16
D18S51	13, 14	14, 14	14, 17	14, 17
D21S11	31.2, 32.2	30, 32.2	30, 30	29, 30
TH01	6, 9	6, 9.3	5, 6	6, 9.3
D3S1358	16, 17	16, 17	15, 15	15, 17
Penta D	11, 14	11, 11	11, 14	11, 12
CSF1PO	8, 12	8, 12	8, 11	12, 12
D16S539	11, 13	9, 11	9, 13	9, 10
D7S820	9, 10	8, 9	8, 8	8, 9
D13S317	11, 12	10, 11	7, 12	10, 13
D5S818	7, 8	7, 11	11, 12	11, 12
Amelogenin	XX	XX	XY	XY

23. DNA thermal cycler

24. Microcentrifuge

25. Microwave or hot plate

Procedure

To prevent cross-contamination, the use of disposable gloves and aerosol-resistant pipet tips is highly recommended.

1. Refer to Exercise 7 for the methods and/or steps used in the collection and concentration of cells and for methods outlining cell lysis and the collection of DNA.

2. Refer to Exercise 7 for the methods and/or steps used in setting up the PCR amplification and for the visualization of STR PCR products.

A Case Study

In 2007, a pregnant 15-year-old female claimed that her father had sexually assaulted her, which resulted in her pregnancy. The father, in his early 50s, was arrested and charged with incestuous pedophilia and sexual abuse. The father denied all claims and stated that his daughter was sexually active with her 18-year-old boyfriend. To establish a genetic profile of the fetus and to determine the biological father, a chorionic villus sample (CVS) from the fetus was collected. Samples (i.e., buccal swabs) were also collected from the mother (i.e., the 15-year-old daughter), the 18-year-old boyfriend, and the alleged father. All samples were sent to the state laboratory for DNA analysis. DNA was extracted from all samples, purified, and subjected to STR analysis using the PowerPlex 16 BIO typing system (Promega Corporation). The fluorescently labeled STR products were then separated by gel electrophoresis, and the DNA molecules captured using a fluorescent detection system. The STR typing results of the mother, the child, the 18-year old boyfriend, and the alleged father are shown in Table 7.

Interpreting Your STR Typing Results

Your group will work together to interpret the data generated from the STR analysis for the above-referenced case study. To determine if the alleged father or the 18-year-old boyfriend is the biological father, you should first determine which alleles (or STR fragments) of the child were donated by the mother. This analysis should be performed for each locus. Second, determine if the remaining allele at each locus of the child could have been contributed by the alleged father or the boyfriend. If one allele at any locus does not match, the alleged father or the boyfriend is excluded. If all of the remaining alleles at each locus "match" the alleged father's or the boyfriend's STR profile, then one is presumed to be the biological father. On the basis of these results obtained from all genetic systems tested, the alleged father or the boyfriend cannot be excluded as the biological father of the child. The paternity probability, determined to be 99.99%, would also support these findings that the alleged father or the boyfriend is, in fact, the biological father.

Questions

1. Based on your analysis of the STR typing results, was the alleged father "excluded" as the biological father, or were his genetic markers consistent with those observed in the child and thus he "could not be ruled out" as the biological father?

2. Based on your analysis of the STR typing results, was the boyfriend "excluded" as the biological father, or were his genetic markers consistent with those observed in the child and thus he "could not be ruled out" as the biological father?

3. Using the same STR typing results, would you be able to determine maternity in this instance? Why or why not?

4. Assuming one of the biological parents' profile was unavailable, would you be able to determine the other parent's genetic contribution to the child? How?

5. Assuming that an alleged father (or the boyfriend) refused to provide a sample for DNA testing but you had legal access to his home, what samples would you collect for analysis? The idea is to collect enough material to generate a DNA profile.

6. If the child displayed homozygosity at one locus (e.g., for FGA: 22, 22), would you be able to assign each allele to either parent? Why or why not?

7. How accurate is DNA paternity testing? Are the results conclusive?

8. Will the DNA paternity test results stand up in court? Explain your answer.

9. The collection of cheek cells (e.g., a buccal swab) is often performed instead of collecting blood as a source of biological material for DNA paternity typing. Will the resulting DNA profile from the buccal swab be as accurate as one degenerated from blood cells? Explain your answer.

13

Experiment 10
Polymerase Chain Reaction (PCR)–Based Tests: Y-Chromosome Short Tandem Repeat (Y-STR) Analysis (a Case Study)

Introduction

The ability to designate whether a sample originated from a male or female contributor is extremely valuable in sexual assault cases as well as in other capital cases. The most popular method for sex typing is the amelogenin typing system because the DNA encoding gender can be amplified in conjunction with STR analysis. However, in some instances, STR and amelogenin analysis is not adequate when multiple males contribute to an evidentiary sample (e.g., a blood sample containing DNA from more than one male). Recently, Y-STR analysis has become available to the forensic community and has provided identification where STR analysis was not definitive.

Several genetic markers have been identified on the Y chromosome that are distinct from markers on the autosomes and are useful for human (male) identification (Table 8). The Y-STR markers are found on the noncoding region located on both arms (i.e., p and q) of the Y chromosome. The Y-STR markers produce a haplotype profile when amplified from male DNA. Such a profile simplifies the interpretation of a mixture containing both a male and female contributor by eliminating the female contribution from the amplification profile. This also eliminates the need to separate semen and vaginal epithelial cells prior to analysis. The Y-STR markers are extremely valuable in sexual assault cases where samples contain multiple male contributors.

Y-STR markers are also useful in the analysis of lineage and the reconstruction of family relationships. In essence, a sample from a male may be compared with those of another male and/or his brother, father, paternal grandfather, or paternal uncles for identification purposes and familiar relationships. Because these markers are only paternally inherited, they are useful in paternity-related matters. In addition, Y-STR markers' use and effectiveness in lineage studies can extend to answering questions of common ancestral geographical origin. Y-STR markers, together with mitochondrial DNA (mtDNA) markers (see Exercise 11), will complement each other in these ancestral analyses.

TABLE 8
Y-STR Loci of the PowerPlex Y System

Y-STR Locus	Sequence
DYS391	TCTA
DYS389I	[TCTG][TCTA]
DYS439	GATA
DYS389II	[TCTG][TCTA]
DYS393	AGAT
DYS390	[TCTG][TCTA]
DYS385a/b	GAAA
DYS438	TTTTC
DYS437	[TCTA][TCTG]
DYS19	TAGA
DYS392	TAT

Objective

In this exercise, DNA has been extracted from male buccal swabs and/or from a human cell line of male origin, incubated with appropriate PCR reagents, and multiple Y-STR loci amplified using the PowerPlex Y System (Promega Corporation). Following PCR, the amplified Y-STR products can be separated and analyzed using polyacrylamide gel electrophoresis or capillary electrophoresis (CE; Figure 13). In this exercise, you will be analyzing data, specific to a case study, generated using CE. As the amplified Y-STR products are separated, the fluorescently labeled DNA molecules are excited by the laser light source, the emission is captured by a detection system, and it is recorded as an electropherogram by the computer.

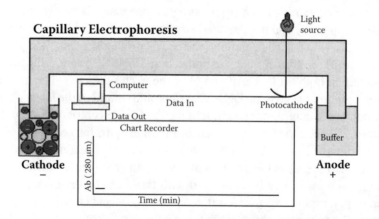

FIGURE 13

Schematic of the Capillary Electrophoresis System. *Note*: Samples are injected into the tube on the left (cathode) and "travel" to the right or to the anode. The fluorescently labeled DNA molecules (or amplified products) are excited by the light source, captured by a detection system, and recorded as an electropherogram by the computer. *Source*: Courtesy of http://www.CEandCEC.com.

Equipment and Material

1. 0.5, 1.0, and 1.5 ml Eppendorf or microcentrifuge tubes
2. 30 or 50 ml conical tubes
3. Mineral oil (optional)
4. 15 ml polypropylene test tube
5. Double-distilled water
6. Ice in buckets
7. Adjustable-volume digital micropipets (2–200 μl range)
8. Aerosol-resistant pipet tips
9. GenePrint STR Systems CSF1PO, TPOX, and TH01 (Promega Corporation, Madison, Wisconsin)
10. Taq DNA polymerase (not supplied in kit)
11. Disposable gloves
12. Agarose
13. Genomic DNA from human cell lines
    A. HEP G2: hepatocellular carcinoma (liver), male (10 ng/μl)
14. Molecular weight markers (526 to 22,621 bp)
15. Ethidium bromide or coomassie blue
16. Bromophenol blue tracking dye
17. 125 ml Erlenmeyer flask
18. TBE buffer (loading buffer)
19. TAE buffer
20. Incubator or water bath at 56°C
21. Electrophoresis systems (gel tray or combs)
22. Power pack or supply
23. Microwave or hot plate
24. DNA thermal cycler
25. Microcentrifuge
26. Tabletop clinical centrifuge

Procedure

To prevent cross-contamination, the use of disposable gloves and aerosol-resistant pipet tips is highly recommended. A helpful organizational sheet is provided at the end of the exercise to record your Y-STR typing results.

1. Refer to Exercise 1 for the methods and/or steps used in the collection and concentration of cells and for methods outlining cell lysis and the collection of DNA.

Setting Up the PCR Amplification

1. Refer to Exercise 7 for setting up the PCR amplification reaction. Keep all samples and reagents on ice.

2. Determine the number of reactions to be set up. Positive and negative controls should also be included when determining the number of reactions.

3. For each reaction, label one sterile 0.5 ml microcentrifuge tube, and place into a rack.

4. To determine the final volume of the master mix needed for all of the reactions, calculate the required amount of each component of the PCR master mix (see Table 9). Multiply the volume (μl) per sample by the total number of reactions (from Step 2) to obtain the final volume (μl).

5. The reaction tubes were placed in a thermal cycler programmed to run in two phases—10 cycles at set parameters, followed by 22 cycles.

For 10 cycles:

Step 1	11.0 min	95°C
Step 2	1.0 min	96°C
Step 3	1.0 min	94°C
Step 4	1.0 min	60°C
Step 5	1.5 min	70°C

Then, for 22 cycles:

Step 6	1.0 min	90°C
Step 7	1.0 min	58°C
Step 8	1.5 min	70°C

Then:

| Step 9 | 30 min | 60°C |
| Step 10 | Soak | 4°C |

TABLE 9
Master Mix for the PowerPlex Y System

PCR Master Mix Component	Volume per Sample (μl)	Number of Reactions	Final Volume (μl)
Sterile Nuclease-free Water			
Gold ST*R 10X Buffer (Promega Corporation)	2.50		
PowerPlex Y 10X Primer Pair Mix	2.50		
AmpliTaq Gold DNA Polymerase (at 5u/μl)*	0.55 (2.75 units)		
Total volume	**25.00**		

Note: Template DNA volume (0.25–1 ng)—up to 19.45 μl.

Note: The parameters outlined above may vary according to the thermal cycler used for the PCR amplification.

Detection of the Y-STR PCR Products

1. Following PCR amplification, the fluorescently labeled Y-STR alleles are separated and sized using polyacrylamide gel electrophoresis (PAGE) or capillary electrophoresis (CE).

2. Samples are denatured by heating at 95°C for 3 min (PAGE) or by diluting with a denaturant solution (CE), then are immediately chilled on ice.

3. The samples are loaded onto a gel for PAGE analysis or introduced into the capillary for CE analysis by injection. For PAGE analysis, multiple samples are separated and analyzed in 2.5 to 3 hr. For CE analysis, only one sample is injected into a capillary tube for separation and analysis; however, this process is completed in a matter of a few minutes.

4. Detection of the sample analyzed by PAGE is performed by scanning each lane and then imaged using a computer detection system. Detection of the sample is performed automatically by the CE instruments by measuring the time span from injection to sample detection with a laser near the end of the capillary. In both instances (i.e., PAGE and CE), the laser excites the fluorescently labeled DNA fragments, which causes a fluorescent light emission. This emission is captured by the detection system and plotted as a function of the relative fluorescence intensity observed from each fluorescent dye attached to the DNA molecule. These signals, recorded as bands on a gel or as an electropherogram for CE, can then be used to detect and quantify the Y-STR PCR products (Figure 14).

Case Study

A woman was walking to her car, which was parked in an underground parking garage. As she was unlocking her car door, a man approached her from behind, forced her into the back seat of her car, and raped her. After the attacker fled the scene, she immediately called the police from her cell phone. The police took her to the local hospital, where she was examined by a sexual assault nurse examiner. Vaginal swabs and a reference sample were collected and sent to the state's forensic laboratory for analysis. The woman had described her attacker as a tall and thin African American male with a tattoo on his right hand (the only part of him that was visible to her during the attack). From that description and a file of known sexual offenders, the police arrested a male suspect. A blood sample was collected from the suspect and sent to the forensic laboratory for analysis. Because the woman was not married, did not have a boyfriend, and had not had consensual sex in several weeks, samples from consensual partners were not needed for analysis.

At the laboratory, the samples were subjected to DNA typing, specifically Y-STR analysis. The vaginal swabs taken from the victim were found to contain sperm and her own cells. The sperm cells were first separated from the victim's epithelial cells, and the DNA isolated using differential extraction. The purified DNA was amplified by PCR and analyzed at 11 Y-STR loci (DYS391, DYS389I, DYS439, DYS389II, DYS438, DYS437, DYS19, DYS392, DYS393, DYS390, and DYS385). The suspect's DNA was also analyzed at the same Y-STR loci.

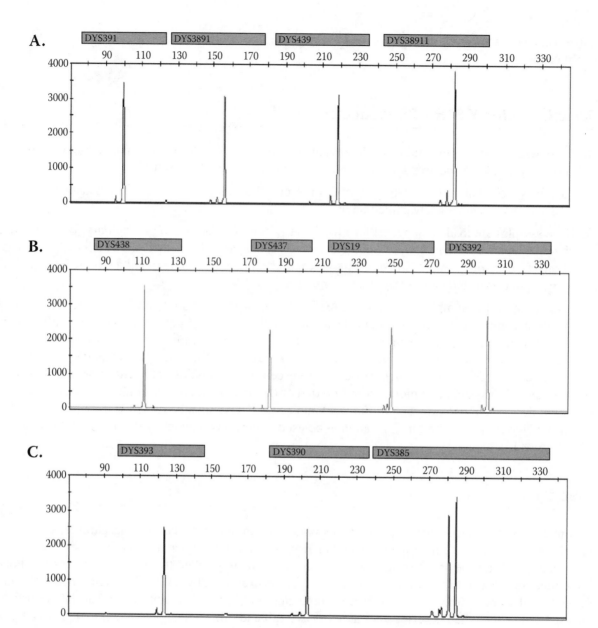

FIGURE 14

The PowerPlex Y System. **Note:** A single source sample from a male contributor was amplified using the PowerPlex Y System. The amplified products were captured using the Applied Biosystems 3130 Genetic Analyzer and analyzed using the GeneMapper ID software to generate the Y-STR profile. Panel A: An electropherogram of the DYS391, DYS389I, DYS439, and DYS389II loci. Panel B: An electropherogram of the DYS438, DYS437, DYS19, and DYS392 loci. Panel C: An electropherogram of DYS393, DYS390, and DYS385. *Source*: Courtesy of the Promega Corporation.

Data Analysis

1. Following amplification, the amplified Y-STR products from the suspect's known reference sample (see Table 10 for Y-STR results) and the sperm fraction from the vaginal swab were separated by CE (Figure 14).

2. The suspect's allelic response at each Y-STR locus can be identified by direct comparison between the Y-STR profile from the sperm fraction (Figure 15) and the allelic ladders (Figure 16). The allelic

TABLE 10
Y-STR Typing Results

Y-STR Loci	Suspect's Reference Sample
DYS391	10
DYS389I	14
DYS439	11
DYS389II	32
DYS438	10
DYS437	14
DYS19	15
DYS392	12
DYS393	14
DYS390	23
DYS385	15, 17

responses observed for each locus should allow for the numerical assignment of each allele. The electropherogram of the negative control should be devoid of any amplification products (results not shown). Conversely, the electropherogram of the positive control should consist of the male DNA standard with known allelic responses (data not shown).

Interpreting Your Y-STR Results

Your group will work together to interpret the electropherograms in Figure 15, Figure 16, and Figure 17. Using the Y-STR typing results from the suspect's reference sample (Table 10), the Y-STR profile from the sperm fraction of the vaginal swab from the victim (Figure 15) generated by CE, and the allelic ladder (Figure 16), determine the allelic numerical designation for each locus analyzed. In addition, determine the overall Y-STR profile from the sperm fraction, and record your observation in the "Y-STR Typing Results" table (below).

A.

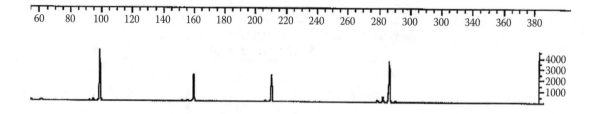

B.

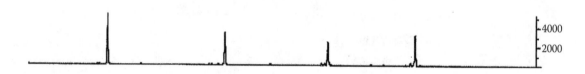

C.

D.

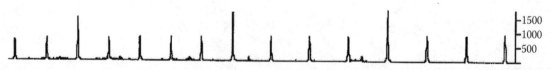

FIGURE 15

PowerPlex Y-STR Typing Results. **Note:** The Y-STR profile, shown in the electropherogram below, was obtained from the sperm fraction of the vaginal swab collected from the victim. Eleven Y-STR loci, specific to the male chromosome, were amplified and separated by capillary electrophoresis. Panel A: An electropherogram of the DYS391, DYS389I, DYS439, and DYS389II loci. Panel B: An electropherogram of the DYS438, DYS437, DYS19, and DYS392 loci. Panel C: An electropherogram of DYS393, DYS390, and DYS385. Panel D: An electropherogram showing the fragments of the Internal Lane Standard.

A.

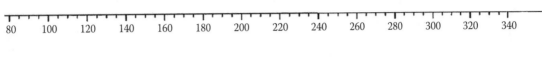
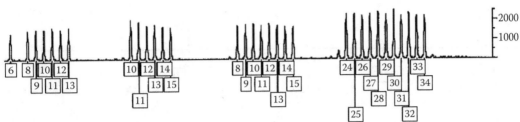

B.

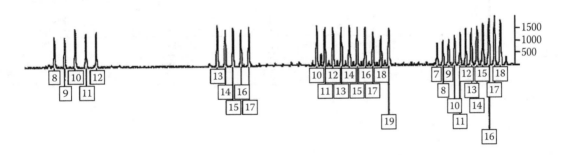

C.

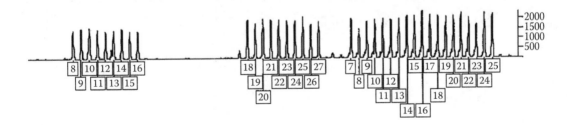

D.

FIGURE 16

The PowerPlex Y Allelic Ladder Mix. **Note:** The allelic components and their allelic or numerical designations. Panel A: An electropherogram of the DYS391, DYS389I, DYS439, and DYS389II loci. Panel B: An electropherogram of the DYS438, DYS437, DYS19, and DYS392 loci. Panel C: An electropherogram of DYS393, DYS390, and DYS385. Panel D: An electropherogram showing the fragments of the Internal Lane Standard.

A.

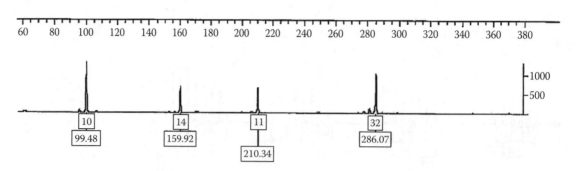

B.

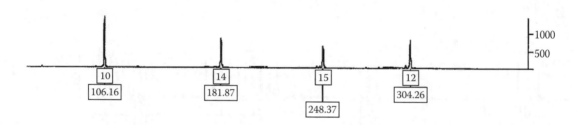

C.

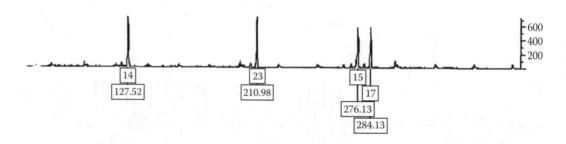

D.

FIGURE 17

The PowerPlex Y-STR Typing Results: Alleles Identified. **Note:** The Y-STR profile, shown in the electropherogram below, was obtained following amplification of the STR loci specific to the male chromosome. Direct comparison between the allelic ladder (Figure 12) and the amplified sperm fraction of the same locus allowed for the numerical assignment of each allele. Panel A: An electropherogram of the DYS391, DYS389I, DYS439, and DYS389II loci. Panel B: An electropherogram of the DYS438, DYS437, DYS19, and DYS392 loci. Panel C: An electropherogram of DYS393, DYS390, and DYS385. Panel D: An electropherogram showing the fragments of the Internal Lane Standard.

Y-STR Typing Results

Y-STR Loci	Sperm Fraction from Vaginal Swab	Suspect's Reference Sample
DYS391		10
DYS389I		14
DYS439		11
DYS389II		32
DYS438		10
DYS437		14
DYS19		15
DYS392		12
DYS393		14
DYS390		23
DYS385		15, 17

Questions

1. Are the Y-STR profiles similar between the sperm fraction and the suspect's known reference sample? Explain your answer.

2. In this case study, the sperm cells were separated from the victim's epithelial cells at the start of the Y-STR analysis. Was this step necessary? Why or why not?

3. How does the Y-STR profile, generated by your group from Figure 15, compare to the Y-STR profile observed in Figure 17?

4. If multiple males were involved in the sexual assault, how would you differentiate between each contributor using Y-STR analysis? What type of Y-STR allelic response would you expect to see at each locus?

5. Would STR analysis complement the Y-STR results? Explain your answer.

14

Exercise 11
Mitochondrial DNA (mtDNA) Analysis

Introduction

Mitochondrial DNA (mtDNA) typing is increasingly used in human identity testing when biological evidence may be degraded, when quantities of the samples in question are limited, or when nuclear DNA typing is not an option. Biological sources of mtDNA include hairs, bones, and teeth. In humans, mtDNA is inherited strictly from the mother. Consequently, mtDNA analysis cannot discriminate between maternally related individuals (e.g., mother and daughter, or brother and sister). However, this unique characteristic of mtDNA is beneficial for missing person cases when mtDNA samples can be compared to samples provided by the maternal relative of the missing person.

In humans, the mtDNA genome is approximately 16,569 bases (A, T, G, and C) in length, containing a "control region" with two highly polymorphic regions (Figure 18, at the top of the figure between the "F" and "P" sites). These two regions, termed Hypervariable Region 1 (HV1) and Hypervariable Region 2 (HV2), are 342 and 268 base pairs (bp) in length, respectively, and are highly variable within the human population. This sequence (the specific order of bases along a DNA strand) variability in either region provides an attractive target for forensic identification studies. Moreover, because human cells contain several hundred copies of mtDNA, substantially more template DNA is available for amplification using PCR than with nuclear DNA.

Mitochondrial DNA typing begins with the extraction of mtDNA from the mitochondria of human cells followed by PCR amplification of the hypervariable regions. The amplified mtDNA is purified, then subjected to the dideoxy terminator method of sequencing (Sanger et al., 1977), with the final products containing a fluorescently labeled base at the end position. The products from the sequencing reaction are separated, based on their length, by gel electrophoresis. The resulting sequences or profiles are then compared to sequences of a known reference sample to determine differences and similarities between samples. Samples are not excluded as originating from the same source if each base (A, T, G, or C) at every position along the hypervariable regions is similar. This sequence, if determined to be

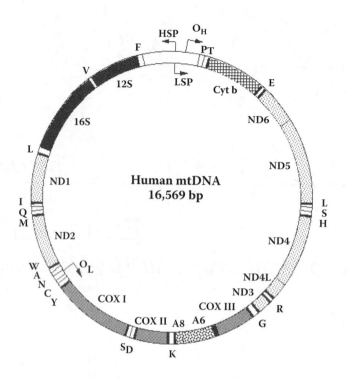

FIGURE 18

Mitochondrial Genome in Humans. **Note:** The two noncoding hypervariable control regions (HV1 and HV2), located within the "D-loop" (between the "F" and "P" sites) of the mtDNA genome, are positioned at the top of the figure. Each of the hypervariable regions is approximately 300 bp in length. The HV1 extends from nucleotide 16024 to 16383, and the HV2 from nucleotide 57 to 372 (Butler and Levin, 1998). *Source*: Courtesy of Columbia University.

similar between a known reference sample and an evidentiary sample, can be entered and searched in a database containing mtDNA sequences from four main racial groups (Caucasians, African Americans, Hispanics, and Asians). The search will generate a number that represents the number of observations of that sequence in each racial subgroup within the database. For example, a sequence might be seen 3 times in the database samples of Hispanic descent and not appear in the remaining database subgroups. Or, a sequence may not be observed at all in the database and is reported as not being observed. This number is usually reported as *1 out of 4800 sequences* or *0 out of 4800 sequences*. However, due to the size of the mtDNA database and to the unknown number of mtDNA sequences in the human population, a reliable frequency estimate is not provided. Consequently, mtDNA sequencing is becoming known as an exclusionary tool as well as a technique to complement other human identification techniques.

Objective

In this exercise, you will isolate mitochondria from cells, extract mtDNA from the mitochondrial fraction, incubate the isolated mtDNA with appropriate PCR reagents, and amplify the hypervariable regions. The amplified mtDNA is then purified, subjected to sequencing, and separated by gel electrophoresis. You will compare the sequencing data from your known reference and unknown sample to a known standard.

Equipment and Material

1. 0.5, 1.0, and 1.5 ml Eppendorf or microcentrifuge tubes

2. 30 or 50 ml conical tube

3. Mineral oil (optional)

4. 15 ml polypropylene test tube

5. Double-distilled water

6. 10% ammonium persulfate solution

7. Adjustable-volume digital micropipets (2–200 µl range)

8. Aerosol-resistant pipet tips

9. Taq DNA polymerase

10. Disposable gloves

11. Ice in buckets

12. mtDNA from human cell lines

    A. HEP G2: hepatocellular carcinoma (liver), male

13. Molecular weight markers (526 to 22,621 bp)

14. mtDNA primers

15. Ethidium bromide and coomassie blue

16. Bromophenol blue tracking dye

17. Tetramethylethylenediamine (TEMED)

18. USB Thermo Sequenase Cycle Sequencing Kit (USB Corporation, Cleveland, Ohio)

19. 20 and 60 cc Hamilton syringe and a 14 gauge needle

20. LI-COR Infrared DNA Analyzer (Model 4300) or equivalent

21. Microcentrifuge

22. DNA thermal cycler

23. Microwave or hot plate

24. Incubator or water bath at 56°C and 92°C

Procedure

To prevent cross-contamination, the use of disposable gloves and aerosol-resistant pipet tips is highly recommended. A helpful organizational sheet is provided at the end of the exercise to record your mtDNA typing data.

The protocol provided below is a basic guide to DNA sequencing using the LI-COR Infrared DNA Analyzer (Model 4300). Various aspects of sequencing are discussed (i.e., mtDNA analysis), including template preparation, primers used, reagents, labeled primer sequencing, gel preparation, and data analysis and interpretation.

The LI-COR system (LI-COR Biosciences, Lincoln, NE) detects DNA using infrared (IR) fluorescence. In the dideoxy sequencing reaction, the DNA polymerase incorporates either a nucleotide or a primer labeled with an IRDye™ into a newly synthesized set of chain-terminated complementary strands. The IRDye™-labeled fragments are separated by gel electrophoresis and are detected using a laser that excites the dye on the DNA fragments. The emission or signal is a series of bands displayed on a computer in a "bar code" format similar to an autoradiograph. The bar code image is captured by the DNA sequencer and analyzed using specific software (e.g., e-Seq). The sequence data are determined for each lane, and the specific order of bases (A, T, G, and C) is determined. The sequence data are presented as a standard chromatogram or as an ASCII text.

A. Template Preparation

The quality and/or purity of the DNA template will dictate the sequence data quality. Several DNA extraction and purification methods were described in previous exercises that will ensure and maximize the quality and purity of the DNA template (see Exercises 1, 2, and 3). In addition to the quality of the DNA, it is important to determine the quantity of DNA in the known and/or evidentiary sample to be analyzed. Similar amounts of DNA template in each reaction will provide consistent data and similar band intensities. Template DNA concentration should range between 0.5 and 1.0 µg/µl (see Tables 11 and 12). The amount of template DNA used in the reaction is based on the size of the DNA sequence between the two primers. If lower yields are obtained, concentrate the DNA by resuspending the final pellet in a smaller volume of buffer (see Exercise 2).

TABLE 11
Template Amount Used in Simultaneous Bidirectional Sequencing (SBS)

Size (bp)	Template (fmoles)
300 – 600	50 – 100
600 – 1200	125 – 225
1300 – 1800	250 – 300
> 1800	300 – 500

TABLE 12
Concentration of Template DNA Recommended for Labeled Primer Cycle Sequencing

Template	Amount (fmoles)
Plasmid	200 – 500
PCR products	20 – 50
M13	100 – 200
Cosmids	50 (1.5 µg)

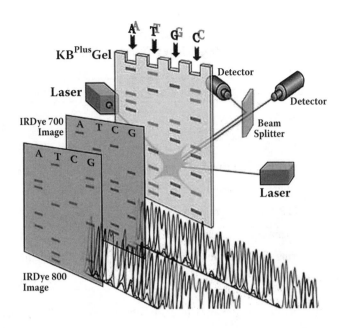

FIGURE 19

The LI-COR Infrared DNA Analyzer (Model 4300). **Note:** Samples are loaded on the polyacrylamide gel and separated by electrophoresis. As the samples pass in front of the scanning laser or microscope, two photodiodes (e.g., the detectors) detect fluorescence. Each detector measures fluorescence from only one of the infrared dyes. A separate image (similar to an autoradiogram) for each IR dye is collected in real time and can be displayed in an Internet browser or in LI-COR application software (e.g., e-Seq or Saga). *Source*: Used with permission from LI-COR Biosciences.

B. Primers

Successful sequencing reactions depend on many factors; however, primer design (i.e., having 40–50% GC content, containing a G or C at the 3′ end, and avoiding base repeats greater than 3 bases), primer purity and concentration are the most critical factors. Many primers are commercially available from several distributors and are, in general, of a high quality. If not commercially available, it is essential that the main impurities (e.g., salts or organic groups) that can affect the sequencing reaction have been removed.

C. Labeled Primer Sequencing

Simultaneous bidirectional sequencing (SBS) uses two labeled primers (e.g., forward and reverse primer pairs labeled with different IR dyes) on a single DNA template in a single reaction. An SBS reaction uses equal amounts of labeled primers (IRDye™ 700 and IRDye™ 800) to obtain equal signal strength in both channels. **Note:** The Model 4300 detection system uses two separate lasers and detection that maximizes sequence accuracy (see Figure 19).

D. Setting Up the PCR Amplification

1. Label four 0.5 ml microcentrifuge tubes as follows: A, T, G, and C.
2. Prepare the following "Template–Primer Master Mix" (see Table 13). Add the largest volume first, and then add solutions in descending order based on volume. Mix the components by pipetting.

TABLE 13
Template–Primer Master Mix

Template/Primer Master Mix Components	Volume
Template DNA	0.3 µl (300 ng)
IRDye™ 700 Forward Primer (1.0 pmol/µl)	1.5 µl
IRDye™ 800 Reverse Primer (1.0 pmol/µl)	1.5 µl
Thermo sequenase reaction buffer	2.0 µl
2.5 mM dNTP nucleotide mix	1.0 µl
Thermo sequenase DNA polymerase	2.0 µl
ddH$_2$O to bring final volume to 17.0 µl	__ µl
Total volume	17.0 µl

3. Add 4 µl of the template–primer master mix to each tube labeled A, T, G, and C.

4. Add 4 µl of the A reagent to the tube labeled A, 4 µl of the T reagent to the tube labeled T, and so on (reagents supplied with Sequenase Cycle Sequencing Kit).

5. Add 1 drop of mineral oil to each microcentrifuge tube to prevent evaporation. This step is required for thermal cyclers without heated lids. Close the tubes, and centrifuge briefly (5 sec).

6. Place the reaction tubes into a thermal cycler programmed to run at least 30 cycles with the following parameters:

Step 1	2.0 min	92°C	Denaturation
Step 2	30 sec	92°C	Denaturation
Step 3	30 sec	54°C	Annealing
Step 4	1.0 min	70°C	Extension
Step 5	Repeat Steps 2–4 for a total of 30 cycles.		
Step 6	Soak or hold	4°C	

Note: The parameters outlined above may vary according to the thermal cycler used for the PCR amplification.

7. Start or "run" the PCR incubation reaction.

8. At the completion of the cycling program, add 4 µl of the IR^2 stop solution to each tube.

9. If mineral oil was used, remove the oil from each sample. Denature the samples at 92°C for 3 min. Then place the samples on ice.

E. Gel Electrophoresis

Assembling the Gel Apparatus

Follow the manufacturer's manual and protocol for specific instructions on assembling the electrophoresis apparatus, preparing the gel, pouring the gel, pre-electrophoresis preparation, starting the run, using the e-Seq software, and disassembling and cleaning up the gel apparatus. The protocol outlined below highlights the major steps in the mtDNA sequencing analysis.

1. Assemble the gel sandwich by laying the back plate (Figure 20, #6) down on the bench (gel side up) and placing two spacers (Figure 20, #5) along the edges of the long axis of the glass plate.

2. Place the front plate (Figure 20, #7, gel side down) on top of the bottom plate containing the spacers, making sure the plates are aligned at the bottom.

3. Place the left and right rail assemblies over the long axis of the plate edges (Figure 20, #8 and #9). Tighten the glass clamp knobs on each rail "finger tight."

Gel Preparation

1. The gel and running buffer solutions are prepared from a 10X TBE buffer. Empty the contents of the KB^{Plus}10X TBE package (supplied with the sequencing kit) in a 1 L beaker, and add distilled water to bring the volume up to 800 ml. Stir the solution until all of the solids have gone into the solution.

2. Bring the final volume to 1 L with distilled water. Store at room temperature.

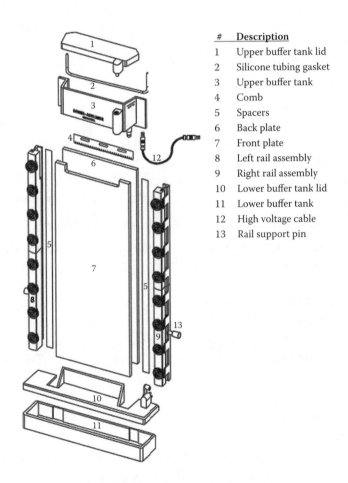

#	Description
1	Upper buffer tank lid
2	Silicone tubing gasket
3	Upper buffer tank
4	Comb
5	Spacers
6	Back plate
7	Front plate
8	Left rail assembly
9	Right rail assembly
10	Lower buffer tank lid
11	Lower buffer tank
12	High voltage cable
13	Rail support pin

FIGURE 20

Expanded View and List of Parts for the Gel Apparatus. **Note:** The KB^{Plus} Gel Matrix is a "ready-to-use solution" containing polyacrylamide. The final gel concentration is 3.7%, that is 66 cm long and 0.2 mm thick. There are other commercial acrylamides available that can be used in the LI-COR system. *Source*: Used with permission from LI-COR Biosciences.

FIGURE 21

Sharkstooth Comb. *Source*: Used with permission by LI-COR Biosciences.

3. Prepare the running buffer (0.8X) by adding 80 ml of the 10X TBE to 920 ml of distilled water, and mix well.

4. For polymerization of the gel, an ammonium persulfate solution (APS) is prepared by adding 0.1 g ammonium persulfate to 1.0 ml of deionized water. The APS should be prepared fresh.

5. Bring 40 ml of the KB^Plus Gel Matrix to room temperature.

6. Add 175 μl of the 10% APS and 17.5 μl of the TEMED to the 40 ml KB^Plus Gel Matrix, and mix thoroughly.

7. Using a 60 cc Hamilton syringe with a 14 gauge needle, "draw" the gel solution (from Step 6, above) into the syringe, and "inject" the solution into the gel cassette.

8. After pouring the gel, invert the sharkstooth comb (Figure 20, #4), and insert it "upside down" at the top of the gel cassette between the front and back plates (Figure 20, #6 and #7).

9. The sharkstooth comb (Figure 21) is inverted prior to polymerization of the gel to make a trough. After polymerization, the comb is removed, inverted "teeth down," and inserted into the gel, forming the wells for the samples.

10. Place the casting plate (part not shown) at the top of the gel cassette and on the front plate. The casting plate will secure the comb until polymerization has occurred.

11. Allow at least 1.5 hr for polymerization.

Electrophoresis Preparation

1. After polymerization, remove the casting plate and the sharkstooth comb.

2. Place the silicone tubing gasket (Figure 20, #2) into the back of the upper buffer tank (Figure 20, #3). Loosen the upper clamp knobs of the rail assembly (Figure 20, #8 and #9), and slide the tank into place; tighten the knobs as before.

3. Open the door of the Model 4300 DNA Analyzer, and place the lower buffer tank (Figure 20, #11) at the base of the unit.

4. Place the gel apparatus on the DNA analyzer (against the heater plate), with the bottom of the gel cassette inside the lower buffer tank. The rail support pins (Figure 20, #13) will hold the gel cassette on the instrument.

5. Fill the upper and lower buffer tanks with the 0.8X TBE running buffer prepared earlier (see "Gel Preparation," Step 3, above). Before adding the running buffer, make sure the drain fitting in the upper buffer tank is closed.

6. Place the upper and lower buffer tank lids (Figure 20, #1 and #10) onto the tanks. Attach the high-voltage cable (Figure 20, #12) to the bottom of the upper buffer tank, and insert the opposing end into the instrument chassis.

Starting the Run

Follow the manufacturer's manual and protocol for specific instructions on starting a new run using the e-Seq software. The e-Seq software automates almost the entire sequencing process by controlling the pre-electrophoretic and electrophoretic runs, and by identifying the bases and their sequence along the mtDNA. After the pre-electrophoretic run, the e-Seq software will automatically pause the process for the user to load samples for analysis. The protocol outlined below highlights the major steps in the mtDNA sequencing analysis.

1. After the "prerun," open the instrument door of the DNA analyzer, and remove the upper buffer tank lid (Figure 20, #1).

2. Using a 20 cc syringe, flush the wells with buffer to remove any debris that may have settled during the pre-electrophoresis run.

3. Load the samples to be analyzed using a Hamilton syringe or a pipet with a flat 0.2 mm micropipet tip. Position the tip between the glass plates, and slowly release the sample into the wells.

4. After loading the samples, replace the upper buffer tank lid, close the instrument door, and push the "Start Run" button.

Base Calling

For base calling and editing the data output, refer to the e-Seq User Guide.

Results

In this exercise, you will be analyzing mtDNA data, specific to a case study, generated using infrared fluorescence detection (Table 14) or data that you have generated using mtDNA isolated from a human liver cell line or from buccal swabs. As the nucleotides are electrophoretically separated, the fluorescently labeled nucleotides (or bases) are excited by the laser light source, the emission is captured by a detection system, and, because of the dual-IR dye capability, the Dual Dye Automated Sequencer permits simultaneous generation of two sequence ladders.

A Case Study

A very respected and dependable young man did not show up for work one day. When the young man did not show up for work on the second day, his employer called his home to see if there was something wrong. When no one answered the phone, the employer tried to contact the young man on his cell phone. The employer was only able to leave a voice mail message. The employer then called the young man's family, only to find out that he had left for work yesterday morning at the usual time.

The family called the police, who, suspecting foul play, launched an investigation. By the end of the week, the young man's body was found in an alley behind the building where he was employed. He had been beaten to death with a blunt-ended object. The victim's body was sent to the medical examiner's office, where his clothing was removed and sent to the state's laboratory for DNA analysis. During the police investigation, a hammer was found in a dumpster in the alley where the body was found. The hammer was placed in a paper bag and sent to the laboratory for analysis. Investigators also determined that the young man was having a relationship with a married woman who also worked for the same employer. When the married woman was questioned, she told the police that her husband had recently learned of the

TABLE 14
Mitochondrial DNA Typing Results

Sample	Hypervariable Region 1 (HV1)													
	16131	16185	16186	16191	16217	16225	16226	16280	16296	16313	16322	16362	16370	16392
Standard	T	C	C	C	T	C	A	C	C	C	A	T	G	T
DD201Q1	•	•	•	•	•	•	G	•	•	G	T	•	•	•
DD201Q2	A	•	•	•	•	•	G	•	•	G	•	•	•	•
DD201Q3	•	•	•	•	•	•	G	•	•	G	T	•	•	R
DD201K1	A	G	G	G	C	A	•	A	A	G	•	A	C	•
DD201K2	•	•	•	•	•	•	G	•	•	G	T	•	•	•

Sample	Hypervariable Region 2 (HV2)													
	75	148	153	155	184	188	191	197	249	265	309.1	315.1	318	
Standard	G	A	A	T	G	A	A	A	A	T	-	-	T	
DD201Q1	C	G	•	•	•	•	•	G	•	C	G	G	•	
DD201Q2	C	G	•	•	•	•	•	G	•	C	•	G	•	
DD201Q3	C	G	•	•	•	•	•	G	•	C	•	G	•	
DD201K1	C	•	T	G	A	T	G	•	T	C	G	G	A	
DD201K2	C	G	•	•	•	•	•	G	•	C	G	G	•	

-: At this position in the published reference sequence referred to as the Anderson sequence ("Standard," top row), there is no nucleotide, and the samples have an insertion.

•: The nucleotide is the same at this position as in the "Standard."

R: Both an A and G were observed at this position.

affair and had vowed to "straighten things out." The police spoke to the husband, who denied any knowledge of the young man's death. Considering all facts, the police arrested the husband and charged him with the murder of the young man. While in custody, the police collected a buccal swab from the husband and sent the sample to the laboratory for analysis. During the autopsy, the medical examiner collected a reference sample from the young man, which was also sent to the laboratory for analysis.

At the laboratory, several strands of hair were discovered on the victim's shirt (designated DD201Q1) and on the "handle" grip of the hammer (designated DD201Q2), and a hair was discovered on the "head" of the hammer (designated DD201Q3). During microscopic examination of the hairs, it was determined that the hair samples were candidates for mtDNA typing. The isolated mtDNA from the hair samples was purified and amplified by PCR, and a complete mtDNA profile obtained comprising the HV1 and HV2 regions. The mtDNA sequence for the evidentiary and known reference samples and the nucleotide substitutions with respect to the standard published reference (i.e., the Anderson sequence) sequence for each sample in this case are presented in Table 14 (Anderson et al., 1981).

Data Analysis

1. Each of the evidentiary hair samples was analyzed according to standard protocol prior to opening or handling the known reference samples. The known samples were also analyzed according to standard protocol established by the laboratory.

2. All negative controls (i.e., reagent blanks and PCR blanks) remained free of contaminating DNA. All positive controls responded as expected.

3. For each sample, a complete mtDNA profile was obtained comprising nucleotide positions 15995 to 16400 (HV1) and from nucleotide 45 to 405 (HV2).

4. The mtDNA profiles of the 3 hairs (DD201Q1, DD201Q2, and DD201Q3) should be compared to the mtDNA profiles of the husband, who is the primary suspect (DD201K1), and the victim (DD201K2). The resulting mtDNA profiles are shown in Table 14.

Interpreting Your mtDNA Results

Your group will work together to interpret the mtDNA profiles shown in Table 14. Compare the mtDNA typing results from the suspect's reference sample (DD201K1) to the mtDNA profiles generated from the hair samples. Also, compare the mtDNA results from the victim's known reference sample to the mtDNA typing results from the 3 hair samples. Each mtDNA profile from the known reference and evidentiary samples should be compared to the standard mtDNA sequence, also known as the Anderson sequence. After completing your analysis, answer the following questions:

1. Is the mtDNA profile of DD201Q2, the hair found on the "handle" grip of the hammer, similar to or different from the mtDNA profile of DD201K2, the victim? Why?

2. Are the victim (DD201K2) and his maternal relatives excluded or included as potential contributors of the hair (DD201Q2)? Why?

3. Are the mtDNA profiles of all 3 hair samples different from or similar to the mtDNA profile of DD201K1, the husband or leading suspect? Why?

4. The Federal Bureau of Investigation (FBI) sponsors an mtDNA population database containing sequences from HV1 and HV2 from several racial groups: Caucasians, Africans, Hispanics, and Asians. The database currently contains over 4800 sequences of North American forensic significance. However, the database is updated frequently and has increased over time. When a sequence from an evidentiary sample is searched in the database, a number will be reported that represents the number of observations of that sequence in each racial subgroup. What is the significance of the statement that 5 out of 4800 sequences were observed when a mtDNA sequence was searched in the database?

Mitochondrial DNA Typing Results

Sample	Hypervariable Region 1 (HV1)											
Standard												

Sample	Hypervariable Region 2 (HV2)											
Standard												

-: At this position in the published reference sequence referred to as the Anderson sequence ("Standard," top row), there is no nucleotide, and the samples have an insertion.

•:The nucleotide is the same at this position as in the "Standard."

R: Both an A and G were observed at this position.

Questions

1. Why were the hair samples subjected to mtDNA typing when the laboratory in this chapter's case study routinely performed STR analysis?

2. Why were the evidentiary samples analyzed separately from the known referenced samples?

3. When the mtDNA profile from the suspect was searched against profiles in the mtDNA database, it was found that the sequence was observed once in the Caucasian database. Because the database includes over 4800 sequences, what is the forensic significance of this frequency? What is the significance of a frequency reported as "not previously observed" in the current database?

4.   When will reliable population frequency estimates for mtDNA types be available?

15

Exercise 12
Assessment of Lumigraph or Autoradiograph Data

Introduction

There are four major steps in the assessment of lumigraph or autoradiograph (autorad) data: 1) visual examination of the lumigraphs or autoradiographs, 2) computer-assisted band size determination, 3) confirmation of visual matcher, and 4) determination of point estimate value.

A. Visual Evaluation of Lumigraphs or Autoradiographs

1. Visually examine the lane(s) of the lumigraph or autoradiograph containing the positive control (K562 DNA). There must be either one or two DNA bands for the K562 DNA positive control, depending on which RFLP or STR locus has been probed or amplified, respectively. If the positive control does not exhibit the expected number of bands for the locus under investigation, the lumigraph or autoradiograph should not be assessed.

2. Visually examine the positive control band(s) (i.e., the K562 DNA bands) for their position relative to the adjacent molecular weight markers. Depending on the locus being probed or amplified, the positive control band(s) should be located in an expected position on the lumigraph or autoradiograph. If the positive control band(s) are not located in the visually expected position on the film, the lumigraph or autoradiograph should not be assessed.

3. Visually examine the lanes of the lumigraph or autoradiograph containing the molecular weight markers. The bands in these lanes must be of sufficient intensity to be used as molecular weight references for the positive allelic control (K562 DNA), the known or reference sample(s), and the questioned or

evidentiary DNA bands. If portions of the molecular weight markers (i.e., the ladder lanes) are not visible, the size of the evidentiary DNA bands cannot be determined in these regions.

4. To assess the quality of the DNA bands, visually examine the lanes of the lumigraph or autoradiograph containing the known or evidentiary DNA. DNA band irregularities in these lanes, such as increased band width (extremely broad bands) or "smiles" (pronounced band curvature), usually indicate potential mobility shifts during electrophoresis and will often compromise the interpretation of the data. If any DNA band for the known or reference sample(s) or the questioned or evidentiary DNA is observed at a position that indicates a molecular weight greater than 10,000 bp, the evaluation of that sample at that locus is considered inconclusive.

5. Based on the assessments of the quality and position of the DNA from the lumigraph or autoradiograph, decide which samples will be subjected to the computer-assisted band-sizing procedure.

B. Computer-Assisted Band Size Determination

The molecular weight determination of each DNA band is carried out using Windows-based or MS DOS–based computer programs (e.g., GenoTyper, GeneScan Analysis, Gel-Pro, and DNA IMAGE ANALYSIS). The forensic DNA analyst is guided through the imaging and sizing procedures by text display on the computer screen. The computer software program enables an objective estimation of the sizes of the DNA fragments in the positive control, the known samples, and each evidentiary sample. The sizing program ends by printing out the calculated fragment sizes or molecular weight, in base pairs, for each of the samples and the allelic control sample. If the DNA fragment sizes for the positive control (HaeIII-digested or amplified K562 DNA) in a particular autoradiograph or a lumigraph are not within the acceptable size range, the lumigraph or autoradiograph in question should not be used for any conclusive match determinations.

C. Confirmation of Visual Matcher

For autoradiographs containing RFLP data, visual matches must be confirmed or rejected through application of the appropriate mathematical procedures. To make a comparison between the test samples (or crime scene samples) and the known reference samples, other than a visual comparison, a quantitative measurement of the DNA fragments observed following hybridization or amplification and visualization needs to be created. Therefore, to accomplish this procedure in the absence of a computer program, the following steps can be carried out manually to determine the molecular weight of each DNA fragment.

1. For each DNA band in the known sample, calculate a value that is 2.5% of the base pair size determined by the sizing procedure. Add the calculated value to the base pair size of the DNA fragment. Also, subtract the calculated value from the base pair size of the fragment.

2. For each fragment in a questioned specimen that has been determined to be presumptively equal in size to a fragment in a known specimen, calculate a value that is 2.5% of the base pair size of the questioned fragment determined by the sizing procedure. Add the calculated value to the base pair size of the fragment. Also, subtract the calculated value from the base pair size of the fragment.

3. Compare the calculated ranges of base pair values for the known and questioned sample bands. If these ranges overlap, the presumptive equality has been confirmed. If the ranges do not overlap, the presumptive equality of the fragment sizes is either inconclusive or exclusionary.

D. Determination of Point Estimate Value

If sample fragment size equalities have been confirmed, the best estimate values may be determined for the fragment bands in the appropriate sample by using the appropriate computer software program.

Semi-Log Graph Paper

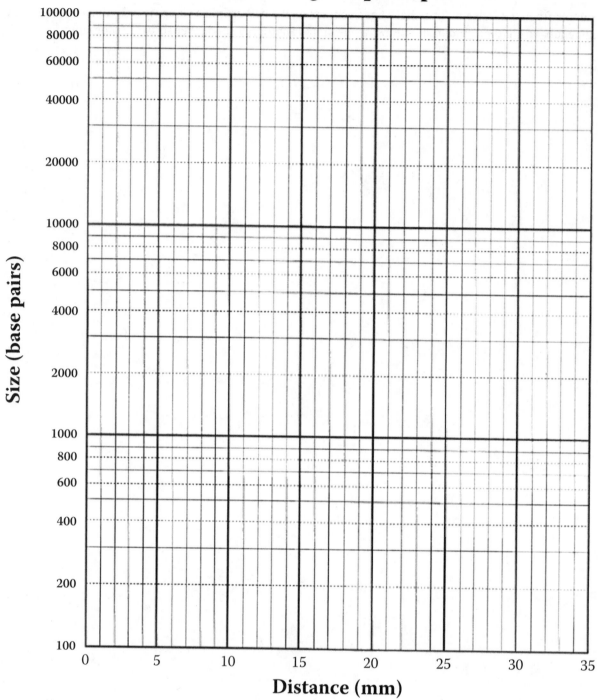

Questions

1. What are the four major steps in the assessment of a DNA profile on a lumigraph or an autoradiograph?

2. When the positive control fails to exhibit the expected number of DNA fragments (or bands) in a lumigraph or an autoradiograph, are the results of the other samples tested (i.e., known or evidentiary) interpretable? Reliable? Why or why not?

3. If the ladder alleles or molecular weight markers are not visible or the DNA bands in these lanes are very low in intensity, the lumigraph or the autoradiograph cannot be assessed. Why?

Appendix A

Composition of Buffers and Solutions

ammonium acetate 7 M: Dissolve anhydrous NH_4OAc in water, and bring the final volume to 100 ml with water (add only 1/2 volume of water to dissolve the NH_4OAc). Sterilize using a sterile 0.45 μm filter. (Expiration: 2 months.)

ammonium persulfate solution (10%): Add 0.1 g ammonium persulfate to 1.0 ml of deionized water. Prepare fresh before use.

analytical gel visual markers: KpnI-digested Adenovirus II DNA. Fragment lengths (in base pairs): 1086, 1699, 2049, 2339, 3648, 5167, 5758, 6478, and 7713.

blotting pads (autoradiographic detection): 11 × 12.5 cm blotting pads.

blotting pads (chemiluminescent detection): 11 × 16 cm blotting pads.

Chelex: Weigh out 1 g of Chelex 100 (100–200 mesh, sodium form). Add 50 mM Tris to the 1.0 g of Chelex to make 10 ml of solution. Adjust the pH to 11 using 4 N NaOH. Store at room temperature. (Expiration: 3 months.)

denaturation solution (autoradiographic detection): 0.4 M NaOH. Combine 500 ml of 4 M NaOH and 4.5 l of distilled water.

denaturation solution (chemiluminescent detection): 0.5 M NaOH/1.5 M NaCl. Combine 250 ml of 4.0 M NaOH and 600 ml of 5.0 M NaCl. Bring to a volume of 2.0 l with distilled water.

dithiothreitol (DTT) 1 M: Dissolve 1.54 g of DTT in 100 μl of sterile sodium acetate, pH 5.2, and bring final volume to 10 ml with sterile deionized water. Store in 100 μl aliquots in 1.5 ml microcentrifuge tubes at –20°C. (Expiration: 1 year.)

ethidium bromide (5 mg/ml) (WARNING: mutagenic substance): Dissolve 1.0 g of ethidium bromide in 200 ml of distilled water. Keep bottle wrapped in foil to protect contents from light.

ethylenediaminetetraacetic acid (EDTA) 0.5 M: Add 80 ml of water to 18.62 g of disodium EDTA-$2H_2O$. Slowly add NaOH pellets to lower the pH to 8.0. When fully dissolved, add more NaOH to bring the pH to 8.0. Adjust final volume to 100 ml. Autoclave, and store at room temperature. (Expiration: 6 months.)

1X final wash solution (chemiluminescent detection): 10 mM Tris-HCl, pH 8.6; and 0.15 M NaCl. Prepare 100 ml of 10X Final Wash Buffer and 900 ml of distilled water. Final wash is a room-temperature wash.

10X final wash buffer (chemiluminescent detection): 2.0 10X Final Wash Buffer. Contains 0.1 M Tris-HCl, pH 8.6, and 1.5 M sodium chloride.

HaeIII: Restriction endonuclease used during RFLP analysis. HaeIII recognizes the four base sequence GGCC cleaving between the G and C to create a "blunt"-ended DNA strand.

herring sperm DNA (10 mg/ml): Dissolve 0.1 g of herring sperm DNA in distilled water, and bring the final volume to 10 ml with water. (Expiration: 6 months.)

high-stringency wash solution (autoradiographic detection): 0.1X SSC and 0.1% SDS. Prepare 5 ml of 20X SSC and 5 ml of 20% SDS, and bring to a final volume of 1.0 l with distilled water. Heat to 65°C before use.

hybridization solution (chemiluminescent detection): Contains 0.5 M sodium phosphate, pH 7.2; 0.5% (v/v) Tween 20; 1% (w/v) Hammersten Casein; and 0.02% (w/v) sodium azide (ACES 2.0 Hybridization Buffer).

K562 DNA (HaeIII-digested): HaeIII-digested K562 DNA is available commercially at 25 ng/µl; it is used as cell line control for postrestriction digestion and analytical gels.

K562 DNA standard (uncut): The typing grade K562 DNA is available commercially and is diluted to 20 ng/µl prior to use.

lambda DNA (HindIII-digested): Commercially available HindIII-digested DNA that is used as a visual marker for yield and postrestriction endonuclease digestion gels. Fragment lengths in base pairs are 125, 564, 2027, 2322, 4361, 6557, 9416, and 23,130.

lambda DNA (uncut): Commercially available DNA for preparation of yield gel quantitation standards. Length in base pairs is 48,502.

loading buffer: 50% Glycerol (v/v), 0.1% (w/v) Bromophenol blue, and 0.1 M EDTA. Prepare 50 ml of 100% glycerol, 0.1 g of bromophenol blue, and 20 ml of 0.5 M EDTA, and bring to a final volume of 100 ml with TE.

loading buffer CH: 10 mM Tris-Cl, pH 7.5; 10% Glycerol (v/v); 0.02% (w/v) bromophenol blue; and 20 mM EDTA. Prepare 400 µl of 2 M Tris-Cl, pH 7.5; 5 ml of 100% glycerol; 10 mg of bromophenol blue; and 2 ml of 0.5 M EDTA, and bring to a final volume of 50 ml with distilled water. *Note*: Loading buffer CH is used only for dilution of molecular weight markers for chemiluminescent detection.

low-stringency wash solution (autoradiographic detection): 2X SSC and 0.1% SDS. Prepare 100 ml of 20X SSC and 5 ml of 20% SDS, and bring to a final volume of 1.0 l with distilled water.

Lumiphos Plus: Chemiluminescent substrate, prepared ready for use.

membrane rinse solution: 0.2 M Tris, pH 7.5; and 2X SSC. Prepare 100 ml of 2 M Tris-Cl, pH 7.5, and 100 ml of 20X SSC, and bring to a final volume of 1.0 l with distilled water.

molecular weight marker probes (chemiluminescent detection): Contains 100 µl molecular weight marker probes specific for the DNA Analysis Marker and 125 ml Lumiphos Plus (ACES 2.0 Marker Probe).

molecular weight markers (autoradiographic detection): DNA sizing standards that are commercially available; fragments range from 640 base pairs to 23,408 base pairs.

molecular weight markers (chemiluminescent detection): DNA sizing markers that are available commercially; these contain a tube with the molecular weight markers in solution and a tube of loading buffer.

neutralization solution (chemiluminescent detection): 1.0 M Tris-Cl, pH 7.5; and 1.5 M NaCl. Prepare 1.0 l of 2.0 M Tris-Cl, pH 7.5, and 600 ml of 5.0 M NaCl, and bring the final volume to 2.0 l with distilled water.

PEG (50%): Dissolve 50 g of polyethylene glycol (MW 8000) in distilled water, and bring the final volume to 100 ml. PEG dissolves very slowly; allow sufficient time to prepare solution.

phenol/chloroform/isoamyl alcohol (100/100/4): Melt 100 g of phenol at 65°C, and pour into a Bellco glass bottle. Add 200 mg 8-hydroxy-quinoline, and mix the solution thoroughly. Add an equal volume of 1.0 M Tris, pH 7.5; transfer to a separatory funnel; mix; and let the phases separate. Drain the lower phenol layer into the Bellco bottle. Drain the upper aqueous phase into a waste beaker. Add an equal volume of 0.01 M Tris, pH 7.5, to the phenol; transfer to the separatory funnel; and mix. Capture the lower phase in the bottle. Capture the upper phase, and determine its pH. If the upper-phase pH is 7.5, cease equilibration procedures. If the pH is less than 7.5, repeat the extraction(s) with 0.01 M Tris until the pH of the upper phase is 7.5. Combine the equilibrated phenol with a solution composed of 100 ml of chloroform and 4 ml of isoamyl alcohol. Cover the solution with 0.01 M Tris, and store at 4°C. This solution (in the ratio of 25:24:1) is also available commercially from various sources.

phosphate-buffered saline (PBS): An isotonic salt solution frequently used to wash residual growth medium from a cell culture monolayer. 5X PBS (per liter) = 40 g NaCl, 1.0 g KCl, 5.75 g Na_2HPO_4, and 19 g KH_2PO_4; autoclave; dilute aseptically to 1X with sterile H_2O prior to use.

proteinase k (20 mg/ml): Dissolve 500 mg of Proteinase K in a small volume of distilled water, and bring to a final volume of 25 ml with distilled water. Dispense into convenient-size aliquots, and freeze.

sarkosyl (20%): Dissolve 20 g of N-lauroylsarcosine in distilled water, and bring to a final volume of 100 ml. Sterilize by filtration using a sterile 0.45 μm filter.

sodium acetate (2 M): Dissolve 41.02 g of sodium acetate in distilled water, and bring to 200 ml with distilled water. Adjust the pH to 7.0 with concentrated HCl, and adjust final volume to 250 ml with distilled water. Autoclave, and store at room temperature.

sodium chloride (5 M): Dissolve 292.2 g of sodium chloride in distilled water, and adjust final volume to 1.0 l with distilled water. Autoclave and store at room temperature. (Expiration date: 6 months.)

sodium dodecyl sulfate (SDS) (20% [w/v]): Add 200 g of sodium dodecyl sulfate to 700 ml water, and heat to 65°C to dissolve. Bring to a final volume of 1.0 l with distilled water.

sodium hydroxide (0.2 M): Prepare 10 ml of 4 M NaOH and 190 ml of distilled water.

sodium hydroxide (4 M): Dissolve 800 g of sodium hydroxide pellets in about 4.2 l distilled water. **CAUTION**: Heat is generated when adding NaOH pellets. Bring to a final volume of 5.0 l with distilled water. Store at room temperature.

20X SSC: 3 M NaCl and 0.3 M NaCitrate, pH 7.0. Dissolve 175.3 g of NaCl and 88.2 g of $Na_3Citrate-2H_2O$ in 800 ml of distilled water. Adjust the solution to pH 7.0 by the gradual addition of HCl. Bring to a final volume of 1.0 l with distilled water.

For 5.0 l, weigh 876.5 g of NaCl and 441 g of $Na_3Citrate-2H_2O$, and dissolve in 4.0 l of distilled water. Adjust the solution to pH 7.0 by the gradual addition of HCl. Bring to a final volume of 5.0 l with distilled water and autoclave.

Note: 2OX SSC is also commercially available as a ready-to-use solution; 2OX SSC has been validated for use in these protocols.

20X SSPE, pH 7.0: 3.6 M NaCl, 0.2 M NaH_2PO_4, and 20 mM EDTA. Dissolve 210.4 g of NaCl and 24.0 g of NaH_2PO_4 (anhydrous) in 900 ml of distilled water. Titrate to pH 7.0 with NaOH (approximately 40 ml of 4 M NaOH). Add 40 ml of 0.5 M $Na_2EDTA-2H_2O$. Bring to a final volume of 1.0 l with distilled water, and autoclave.

stain extraction buffer: 10 mM Tris-Cl, 0.1 M NaCl, 2% SDS, 10 mM EDTA, and 39 mM DTT. Dissolve 1.21 g of Tris and 5.84 g of NaCl in 500 ml of distilled water. Add 100 ml of 20% SDS and 20 ml of 0.5 M $Na_2EDTA-2H_2O$, and adjust pH to 8.0 with HCl. Bring to a final volume of 1.0 l with distilled water.

Supplement with DTT before use. To 100 ml of the above solution, add 0.6 g of powdered DTT, and stir until dissolved. Store at room temperature. (Expiration date: The final solution is good for no more than 2 weeks.)

1X strip solution (chemiluminescent detection): 10 mM Tris-Cl, pH 7.5; 1 mM EDTA; and 0.5% Tween 20. Prepare 50 ml of 10X Strip Solution and 450 ml of distilled water.

10X strip solution (chemiluminescent detection): 0.1 M Tris-Cl, pH 7.5; 10 mM EDTA; and 5% Tween 20. Prepare 100 ml of 2.0 M Tris-Cl, pH 7.5; 40 ml of 0.5 M EDTA; and 100 ml of 100% Tween 20. Bring to a final volume of 2.0 l with distilled water.

1X TAE: 40 mM Tris-acetate, pH 8.3; and 1 mM EDTA. Prepare 50 ml of 2OX TAE and 950 ml of distilled water.

20X TAE: 0.8 M Tris-acetate, pH 8.3; and 20 mM EDTA. Prepare 96.6 g of Tris base, 22.8 ml of glacial acetic acid, and 40.0 ml of 0.5 M EDTA, pH 8.0. Bring to a final volume of 1.0 l with distilled water, and autoclave.

    For 5.0 l, weigh 483 g of Tris base, 114 ml of glacial acetic acid, and 200 ml of 0.5 M EDTA, pH 8.0. Bring to a final volume of 5.0 l with distilled water.

TE buffer: 10 mM Tris-Cl, pH 7.5; and 0.1 mM EDTA. Prepare 1.21 g of Tris base and 0.037 g of Na_2EDTA. Dissolve Tris in 800 ml of distilled water, and adjust the pH to 7.5 with HCl. Add EDTA, check the pH, and adjust to 7.5 if required. Bring the final volume to 1.0 l with distilled water, and autoclave.

TE-9 buffer: 0.5 M Tris, pH 9.0; 20 mM EDTA; and 10mM NaCl.

TNE (pH 8.0): 10 mM Tris-Cl, pH 7.5; 0.1 M NaCl; and 1 mM EDTA. Prepare 2.5 ml of 2 M Tris-Cl, pH 7.5; 10 ml 5 M NaCl; and 1.0 ml 0.5 M EDTA, pH 8.0. Add distilled water to 400 ml. Titrate to pH 8.0 with 0.1 N NaOH. Bring to a final volume of 500 ml with distilled water and autoclave.

transfer membrane (autoradiographic detection): Biodyne B (Pall Biosupport, Port Washington, New York).

transfer membrane (chemiluminescent detection): Biodyne A (Pall Biosupport, Port Washington, New York).

transfer solution (chemiluminescent detection): 10X SSC. Prepare 1 liter of 20X SSC and 1 liter of distilled water.

transfer sponges (chemiluminescent detection): Lifecodes Corporation, Stamford, CT.

Tris (2 M): Dissolve 242.2 g of Tris base in 800 ml distilled water. Adjust to pH 7.5 with concentrated HCl. Bring the final volume to 1.0 l with distilled water, and autoclave.

Tris-acetate EDTA buffer (TAE buffer): Common electrolyte reagent for the electrophoresis buffer for large (> 12 kb molecular weight) DNA. 50X TAE (per liter) = 242.0 g Tris base; 100 ml 0.5M Na_2EDTA, pH 8.0; and 57.1 ml glacial acetic acid; autoclave. Working concentration is 1X TAE.

Tris borate-EDTA buffer (TBE buffer): Common electrolyte reagent for the electrophoresis buffer for low (< 1 kb) molecular weight DNA. 20X TBE (per liter) = 121 g Tris base, 61.7 g sodium borate, and 7.44 g Na_2EDTA. Working concentration is 1X TBE.

VNTR locus oligonucleotide probes (chemiluminescent detection):

D2S44	Lifecodes Corporation NICE format
D2S44	Promega GenePrint Light (#DK5411)
D10S28	Lifecodes Corporation NICE format
D10S28	Promega GenePrint Light TBQ7 (#DK632A)
D17S79	Promega GenePrint Light D17S79 (#DK5431)
D5S110	BRL ACES™ Probe LH1 (#14232-011)*
D4SI39	BRL ACES™ Probe pH30 (#24230-013)*
D1S7	BRL ACES™ Probe MS1 (#14231-013)*
D1S7	Lifecodes Corporation NICE format

    * Life Technologies, Inc. (Gibco BRL), Rockville, MD.

Wash I Concentrate (chemiluminescent detection): 2.0 Wash Buffer I Concentrate. Contains 0.5 M sodium phosphate, pH 7.2; and 5% (v/v) Tween 20.

0.5X Wash I solution (chemiluminescent detection): 25 mM sodium phosphate, pH 7.2; and 0.25% Tween 20. Prepare 50 ml of Wash I Concentrate and 950 ml of distilled water. Heat solution to 55°C prior to use.

1X Wash I solution (chemiluminescent detection): 50 mM sodium phosphate, pH 7.2; and 0.5% Tween 20. Prepare 100 ml Wash I Concentrate and 900 ml of distilled water. Heat solution to 55°C prior to use.

Using a Micropipet

Introduction

Most chemical reactions in forensic DNA analysis are performed in small volumes of liquid, partly because DNA is available only in small quantities and because reagents and enzymes are expensive. The various reactions are performed in small microcentrifuge tubes (0.2–1.5 ml) in volumes as small as 0.5 μl. Consequently, the forensic DNA analyst must be able to dispense such small volumes correctly and accurately. Dispensing such volumes of liquids is accomplished using micropipets. The series of steps below provides an overview on the structure and use of the micropipet.

Structure of the Micropipet

Take a micropipet in your hand, and become familiar with the various parts (see Figure 22). On top of the micropipet is the plunger button for filling with and dispensing liquids. The second button, the tip ejector button, allows the user to eject the disposable tip, thus eliminating the need to touch the tip and/or liquid. An inset wheel or knob, the volume adjustment dial, permits the user to adjust the volume; whereas the dial with numbers (digital volume indicator) indicates the volume that has been selected. The plastic shaft at the base of the micropipet holds the disposable tip.

The micropipets are available in different ranges or capacities:

- P20: up to 20 μl (up to 0.02 ml)
- P200: 20 to 200 μl (up to 0.2 ml)
- P1000: 200 to 1000 μl (up to 1.0 ml)

Micropipets are expensive and can be easily damaged if not handled properly. When using the micropipet always adhere to the follow rules:

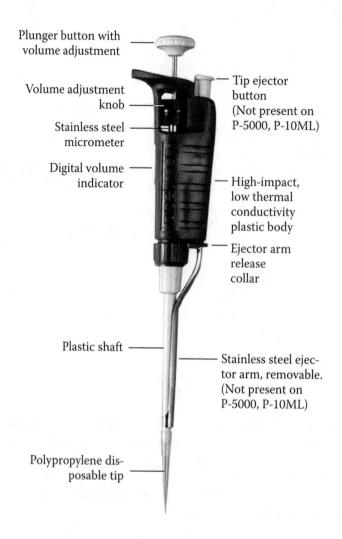

Plunger button with volume adjustment

Volume adjustment knob

Stainless steel micrometer

Digital volume indicator

Tip ejector button (Not present on P-5000, P-10ML)

High-impact, low thermal conductivity plastic body

Ejector arm release collar

Plastic shaft

Stainless steel ejector arm, removable. (Not present on P-5000, P-10ML)

Polypropylene disposable tip

FIGURE 22

A Rainin Classic Pipetman. *Source*: Image used with permission of Rainin Instrument, LLC.

- Never rotate the volume adjustment knob below the lower-volume limit or above the upper-volume limit.

- Never lay the micropipet on the bench; always replace in the stand when not in use. This prevents liquid from entering the pipet and causing damage.

- Never immerse the plastic shaft of the pipet into fluid without a tip in place. Always use a new tip for each different reagent. Use the proper size tip for each pipettor.

- Always slowly release the plunger when withdrawing or dispensing liquids.

    Before using the pipet, familiarize yourself with the feel of the pipet. Hold the pipet in your writing hand. With your thumb, slowly lower and raise the plunger. As you press down on the plunger you will feel resistance at the "first stop," but you can continue to press until the plunger stops. The pipet is filled by pressing the plunger down to the first stop and slowly releasing the plunger. The first stop plus the second stop, or end, will empty the pipet.

Using the Pipet

1. Check the top of the micropipet's plunger button to select the pipet that you will need. Use a pipet with a volume greater than the amount to be pipetted. Refer to the range of sizes above.

2. To select the volume of liquid to be pipetted, rotate the volume adjustment knob until the digital indicator reaches the desired volume.

3. Place a disposable tip on the plastic shaft of the micropipet. Press firmly to ensure that the tip is in place.

4. Press the plunger down to the first stop. Hold the micropipet vertically, and place the disposable tip into the liquid to be pipetted. It is only necessary to place the tip in the liquid to a depth of several millimeters.

5. Slowly release the plunger button to its original position. Make sure the liquid is drawn into the tip.

6. Withdraw the tip from the liquid.

7. To dispense the liquid sample, place the tip against the wall of the receiving tube, and press the plunger down to the first stop, then to the final stop to dispense any remaining liquid.

8. While the plunger is still pushed down, remove the tip from the tube, and allow the plunger to slowly return to its original position.

9. Discard the disposable tip into a waste container by pressing the tip ejector button.

allele: An alternative form of a gene occupying a given location on a chromosome that determines alternative characteristics in inheritance.

allele-specific oligonucleotides (ASO): A short, specific DNA sequence that is used as a probe in the AmpliType PM/DQA1 test to detect a unique sequence.

amino acids: Building blocks of proteins. Each protein consists of a specific sequence of amino acids. There are 20 common amino acid molecules that can make up proteins.

AmpliTaq: Recombinant form of the naturally occurring thermostable DNA polymerase from the organism *Thermus aquaticus*.

anneal: The base pairing of complementary polynucleotides to form a double-stranded molecule.

antiparallel: The manner in which two complementary polynucleotides base-pair to one another; the 5′ and 3′ ends of each molecule are reversed in relation to each other, so that the 5′ end of one strand is aligned with the 3′ end of the other strand. Antiparallel base pairing accompanies the formation of double-stranded DNA (DNA:DNA), double-stranded RNA (RNA:RNA), and DNA–RNA hybrids (DNA:RNA).

autoradiograph (*also* **autoradiogram**): A photographic record of the spatial distribution of radiation in an object or specimen. It is made by placing the object very close to a photographic film or emulsion.

autoradiography: The process by which an autoradiograph is made.

autosome: Chromosomes that are in the same number and kind in males and females; chromosomes other than the sex chromosomes.

avidin (streptavidin) enzyme conjugate (HRP-SA): A nonisotopic detection system used with the AmpliType PM/DQA1 typing system. Biotin is covalently attached to each primer pair. The biotin-labeled amplified products are allowed to hybridize to the DNA probes immobilized on the nylon test strips. The strips are reacted with the enzyme horseradish peroxidase (HRP) covalently bound to streptavidin (SA). This HRP–SA conjugate can bind only to hybridized or double-stranded targets. If hybridization occurs, the HRP–SA conjugate will react with the colorless substrate, causing a blue color to develop. The spot remains colorless if no hybridization occurred, hence a negative response.

bacteriophage lambda: A virus that infects and is propagated in a bacterial host, often for cloning purposes. Among the best characterized and most widely exploited are derivatives of the A bacteriophage.

base: One of five molecules that make up the informational content of DNA and RNA. In DNA, bases pair across the two chains of the double helix: adenine (A) with thymine (T), and guanine (G) with cytosine (C). RNA is single stranded and contains uracil (U) instead of thymine.

base pair: Two complementary nucleotides bonded together at the matching bases (A and T or C and G) to form a double-stranded complex; the length of the DNA is often described in base pairs (bp).

base pairing: The formation of hydrogen bonds between the nitrogenous bases of two nucleic acid molecules.

beta particle: An elementary particle emitted from a nucleus during radioactive decay. It has a single electrical charge. A negatively charged B particle is identical to an electron. A positively charged B particle is known as a positron.

biotechnology: The set of biological techniques developed through basic research and now applied to research and product development; in particular, the use of microorganisms and plant and animal cells to produce useful materials, such as food, medicine, and other chemicals.

biotin: A small vitamin used to label nucleic acids for a variety of purposes, including nonisotopic hybridization by chemiluminescence or chromogenic techniques.

buoyant density: A measure of the ability of a substance to float in a standard fluid. For example, differences in the buoyant densities of RNA and DNA allow them to be separated in a gradient of cesium chloride or cesium trifluoroacetate.

capillary electrophoresis (CE): A technique for separating DNA from a fluid substrate; the sample is injected into a capillary tube, which is then subjected to a high-voltage current that separates its chemical constituents based on charge and size.

cDNA (complementary DNA): DNA synthesized from an RNA template. The single-stranded form of cDNA is an important laboratory tool (e.g., as a probe) for isolating and studying the expression of individual genes.

chemiluminescence: A nonisotopic hybridization detection technique. Chemiluminescence is the production of visible light by a chemical reaction.

chromatin: The complex of genomic DNA and protein found in the nucleus of a cell in interphase.

chromosome jumping: A technique very similar to chromosome walking with the exception that very large fragments of DNA (> 100 kb) under investigation are purified by pulse field gel electrophoresis. Thus, moving from one end of this DNA fragment, for subcloning purposes, constitutes more of a "jump" than a "walk." *See* **chromosome walking**.

chromosome walking: The systematic isolation of a set of clones containing overlapping DNA fragments that collectively makes up a specific genomic region. The process is initiated by identification of a unique site recognized by a sequence-specific probe. When a clone is retrieved from a library, the process of chromosome walking continues by subcloning and rehybridization to the library. In each successive hybridization, the probe corresponds to the 3′ or 5′ end of the clone previously recovered from the library. Thus, the resulting series of overlapping clones permits locus characterization within, upstream of, and downstream of a particular locus. This approach can be exploited to "walk" either toward or away from a locus of interest.

clone (noun): A collection of genetically equivalent cells or molecules.

clone (verb): A series of manipulations designed to isolate and propagate a specific nucleic acid sequence or cell for characterization, storage, or further amplification.

clone bank: Older terminology for what are currently known as cDNA libraries and genomic DNA libraries.

cloning: 1. A collection of genetically equivalent cells or molecules. 2. Asexually producing multiple copies of genetically identical cells or organisms descended from a common ancestor. *Compare with* **gene cloning**.

coding strand: In double-stranded DNA, the strand that has the same sequence as the resulting RNA (except for the substitution of uracil for thymine).

codon: A triplet of nucleotides in an RNA molecule that specifies the placement of an amino acid during protein synthesis.

competitive PCR: A very sensitive method for quantification of transcript abundance by the inclusion of a new DNA sequence, which competes for primers and deoxyribonucleotide triphosphates (dNTPs), in the same reaction tube as the experimental sample. By varying the amount of the competitor DNA (also known as a DNA mimic), a dilution will be identified in which the concentrations of target and mimic are equivalent, allowing for very accurate quantification.

complementary DNA (cDNA): DNA enzymatically synthesized in vitro from an RNA template by reverse transcription. cDNA may be single-stranded or double stranded, as required by the parameters governing a particular assay. The synthesis of cDNA represents a permanent biochemical record of the cellular biochemistry and also provides a means by which that record can be propagated.

constitutive gene expression: Interaction of RNA polymerase with the promoters of specific genes not subjected to additional regulation. Such genes are frequently expressed continually at low or basal levels, and are sometimes referred to as genes with housekeeping functions.

curie (Ci): A basic unit for measuring radioactivity in a sample. One curie is equivalent to 3.7×10^{10} becquerel (i.e., disintegrations per second).

cytoplasm: The cellular contents found between the plasma membrane and the nuclear membrane.

denaturation (of nucleic acids): Conversion of DNA or RNA from a double-stranded form to a single-stranded form. This can mean dissociation of a double-stranded molecule into its two constituent single strands, or the elimination of intramolecular base pairing.

deoxyribonucleic acid (DNA): The substance of heredity; a large linear molecule that consists of deoxyribose sugar, phosphate groups, and the bases adenine, thymine, guanine, and cytosine, and that carries the genetic information that cells need to replicate and to produce proteins. A polymer of deoxyribonucleoside monophosphates, assembled by a DNA polymerase. In vivo, DNA is produced by the process known as replication. DNA can also be synthesized using a variety of in vitro methods, such as the polymerase chain reaction.

differential display PCR (DD-PCR): A method for identification of uniquely transcribed sequences among two or more RNA populations. DD-PCR is a PCR-based method that utilizes large combinations of relatively short primers to ensure amplification of all transcribed RNAs in the form of cDNA. Electrophoretic comparison of the products of each reaction shows products of identical molecular weight when a transcript is common to the biological samples under investigation; a band in only one lane is observed if gene expression has been induced or repressed.

differentiation: The process of biochemical and structural changes by which cells become specialized in form and function.

diploid: Having two complete sets of chromosomes (two of each chromosome). *Compare to* **haploid** *and* **triploid**.

directional cloning: Unidirectional insertion of a DNA molecule into a vector accomplished by placement of different sequences or restriction enzyme sites at the ends of double-stranded cDNA or genomic DNA molecules.

direct repeats: Identical or closely related sequences present in two or more copies in the same orientation on the same molecule of DNA; they are not necessarily adjacent.

DNA (deoxyribonucleic acid): *See* **deoxyribonucleic acid**.

DNA polymerase I: A prokaryotic enzyme capable of synthesizing DNA from a DNA template. The native DNA polymerase I, also known as the holoenzyme or Kornberg enzyme, manifests three distinct activities: 5′ to 3′ polymerase, 5′ to 3′ exonuclease, and 3′ to 5′ exonuclease. *See also* Klenow fragment.

DNA probe: A specific sequence of single-stranded labeled DNA that is used to determine the presence of complementary nucleic acid sequences.

DNA profile: The pattern of band lengths on an autoradiograph (*see* autoradiography) representing all of the tests to link DNA samples with probes.

DNA sequencing: A technology for determining the order of nucleotides in a specific DNA molecule.

dominant: Pertaining to the form of a gene that exerts its effect when present in the individual in just a single copy; an expressed trait.

dot blot analysis: A rapid, quantitative assay for determining the prevalence of a DNA or RNA sequence in a sample. Denatured samples are applied directly to a filter without prior electrophoretic separation. Results are based on signal intensity within the "dot." Dot blot analysis lacks the qualitative component associated with gel electrophoresis. *See also* **slot blot analysis**.

downstream: A term that refers to sequences proceeding further in the direction of expression (i.e., in the 3´ direction); for example, the coding region is downstream from the initiation codon.

duplex: The formation of a double-stranded molecule or portion thereof by the base pairing of two complementary polynucleotides.

electropherogram: A recording of the separated DNA components of a sample produced by gel or capillary electrophoresis. A photograph or printout of a gel or capillary separation made after electrophoresis, which records the spatial distribution of macromolecules within the gel or capillary.

electrophoresis: A type of chromatography in which macromolecules (i.e., proteins and nucleic acids) are resolved through a matrix based on their charge.

ethidium bromide (EtBr): A planar, intercalating agent used to visualize nucleic acids, both DNA and RNA. This dye emits a bright orange fluorescence when UV irradiated; thus, gels that contain samples can be photographed for future reference. Standard ethidium bromide stock solution is 10mg/ml in water; standard staining concentration is 0.5–1.0 mg/ml.

ethylenediamine tetraacetic acid (EDTA): A chemical preservative added to biological samples to inhibit the activity of enzymes that are responsible for degrading DNA.

exon: A portion of a eukaryotic gene represented in the mature mRNA molecule. Exons may or may not be translated.

formaldehyde (HCHO): A commonly used denaturant of RNA.

FTA collection card: An absorbent cellulose-based paper that contains chemical substances to inhibit bacterial growth and to protect the DNA from enzymatic degradation. Liquid samples such as blood and saliva are often collected and "spotted" onto the card for short- or long-term storage at room temperature. FTA is a registered trademark of Flinders Technologies, Pty. Ltd.

gel electrophoresis: In RFLP analysis, the process of separating DNA (cut or uncut) by size in an electrical field; the different-sized fragments move at different rates through the gel.

gene: The fundamental physical and functional unit of heredity. A gene is an ordered sequence of nucleotides located in a particular position on a chromosome that ultimately encodes for the synthesis of a polypeptide.

gene cloning: Isolation and propagation of a gene or gene fragment by inserting it into a suitable vector or host and allowing it to multiply.

gene expression: The manifestation of the genetic material of an organism in the form of a functional polypeptide.

gene library: A collection of DNA fragments (introduced into a virus or plasmid) that, when taken together, represents the total DNA of a certain cell type or organism. An older term for gene library is "clone bank."

gene mapping: Determining the relative locations of different genes on chromosomes.

gene regulation: The processes controlling the synthesis or suppression of gene products.

gene splicing: Joining pieces of DNA from different sources using recombinant DNA technology.

gene therapy: A set of experimental techniques for the introduction of a normal functioning gene into a cell in which the gene is defective.

genetic code: The language in which the instructions of DNA are written. It consists of triplets of nucleotides (codons), with each corresponding to an amino acid or a signal to start or stop protein synthesis.

genome: The entire chromosomal DNA found in a cell; its size is generally given in the total number of base pairs. In some applications, it may be useful to distinguish nuclear genomic DNA from the mitochondrial genome.

genomic DNA: Chromosomal DNA.

genotype: The genetic composition of an individual cell or organism; the total of all the genes present in an individual.

haploid: Having one complete set of chromosomes (one of each chromosome, as found in gametes). *Compare to* **diploid** *and* **triploid**.

haplotype: Refers to the genetic constitution of an individual chromosome. Haplotype may refer to only one locus or to an entire genome. In the case of humans, a genome-wide haplotype comprises one member of the pair of alleles for each.

heterozygous: Both alleles at a given locus on each of a pair of homologous chromosomes are different; one is inherited from each parent.

histones: Proteins that help to organize the chromosomes by their association with DNA. Histone proteins are among the most highly conserved among all eukaryotic genes.

homologous chromosome: Chromosomes that share an identical sequence of genes, but may carry similar or different alleles at the same loci; they associate in pairs.

homozygous: Both alleles are the same at a given locus; one is inherited from each parent.

housekeeping gene: A gene that is expressed, at least theoretically, at constant levels in all cells, the products of which are required to maintain cellular viability. Because of their purported invariance, assay of transcription of these sequences is often performed to demonstrate that an *overall* change in gene expression has *not* occurred, in the context of an experimental manipulation.

human leukocyte antigen (HLA): Antigen (foreign body that stimulates an immune response) located on the surface of most cells (sperm and red blood cells excluded) and different among individuals. HLA DQA1 is a particular class of HLA whose locus has been completely identified (i.e., sequenced) and used in forensic typing.

hybridization: The formation of hydrogen bonds between two nucleic acid molecules that demonstrate some degree of complementarity. The specificity of hybridization is a direct function of the stringency of the system in which the hybridization is being conducted.

hydrogen bonding: The highly directional attraction of an electropositive hydrogen atom to an electronegative atom such as oxygen or nitrogen. This is the manner of interaction between complementary bases during nucleic acid hybridization. *See also* **base pairing**.

image: The document that image analysis software works upon. "Image" may also refer to the original artwork, graphics, or photograph that is scanned or imported into image analysis software.

image analysis: An electronic method for the digital capture and storage of an image, accompanied by the automated measurement of parameters such as molecular weight, mass, relative abundance, and optical density of various objects in the image (e.g., bands on a gel).

intron: A DNA sequence that interrupts the coding sequences (exons) for a gene product. After information from the genes is transcribed into new strands of heterogeneous nuclear RNA (hnRNA), the introns are spliced out of the RNA molecule, and are not represented in the mature mRNA. Although the functions of introns are unknown, it has been postulated that some introns have a role in regulating gene expression.

isotope: One of two or more atoms with the same atomic number but different atomic weights.

Klenow fragment: The large fragment of *E. coli* DNA polymerase I, generated by cleavage of the holoenzyme with subtilisin, or obtained by cloning. The Klenow fragment manifests the 5′ to 3′ polymerase and 3′ to 5′ exonuclease activities but lacks the often troublesome 5′ to 3′ exonuclease activity associated with the intact enzyme. *See also* **DNA Polymerase I**.

library: A collection of clones that partially or completely represent the complexity of genomic DNA or cDNA from a defined biological source, one or several of which are of immediate interest to the investigator. Members of the library, which consists of cDNA or genomic DNA sequences ligated into a suitable vector, may be selected or retrieved from the library by nucleic acid hybridization or, in the case of expression vectors, by antibody recognition.

ligase: An enzyme that catalyzes the formation of the phosphodiester bond between the termini of two DNA molecules.

locus: The precise position of a particular gene, and any possible allele, on a chromosome.

marker: A very generic term that can refer to any allele of interest in an experiment. Also, marker can refer to a molecular standard.

melting temperature: *See* T_m.

messenger RNA (mRNA): The mature product of RNA polymerase II transcription. In eukaryotic cells, mRNA is derived from heterogeneous nuclear RNA (hnRNA) and, in conjunction with the protein translation apparatus, is capable of directing the translation of the encoded polypeptide.

mismatch: One or more nucleotides in a double-stranded molecule that do not base pair. In order for mismatches to be tolerated, the temperature of annealing must be sufficiently below the melting temperature, T_m; at the T_m, only perfectly matched duplexes are stable. The location and context of mismatching have profound ramifications with respect to primer annealing in the polymerase chain reaction.

multiplex PCR: Simultaneous amplification of two or more targets in the same PCR reaction.

Northern analysis (also Northern blotting): A technique for transferring electrophoretically chromatographed RNA from an agarose gel matrix onto a filter paper, for subsequent immobilization and hybridization. The information gained from Northern analysis is used to qualitatively and quantitatively assess the expression of specific genes.

nucleotide: A subunit or molecule of DNA or RNA consisting of a 5-carbon sugar (ribose or deoxyribose), a nitrogenous base (adenine, cytosine, guanine, thymine, or uracil), and a phosphate group. Nucleotides are the building blocks used to assemble both DNA and RNA.

oligonucleotide: A short, artificially synthesized, single-stranded DNA molecule that can function as a nucleic acid probe or a molecular primer. Oligonucleotide can also refer to a short fragment of RNA.

palindrome: A segment of duplex DNA in which the base sequences of the two strands exhibit a twofold rotational symmetry about the central axis. Restriction endonucleases often recognize and cut the DNA at a variety of such palindromic sites.

PCR (polymerase chain reaction): A systematic, primer-mediated enzymatic process for the geometrical amplification of a target DNA sequence. PCR product can be generated from as little as one molecule of target material (DNA or RNA) under optimal conditions.

phenotype: The observable characteristics of an organism or individual.

photodocumentation: A method for preserving the image of a gel immediately after electrophoresis, or after hybridization with a labeled probe. Media that support photodocumentation include Polaroid film, X-ray film, thermal paper, and digital storage. *See* **image analysis**.

plasmid: A covalently closed, double-stranded DNA molecule capable of autonomous replication in a prokaryotic host (eukaryotic plasmids have also been developed). Plasmids can accept foreign DNA inserts, usually less than 10 kb, and often contain a variety of selectable markers and ancillary sequences for characterization of the insert DNA.

polymerase chain reaction (PCR): A primer-mediated enzymatic process for the systematic amplification of minute quantities of specific genomic or cDNA sequences. This technique, which has revolutionized molecular biology, mimics DNA replication. It has the advantage of being a very sensitive technique that can be performed in a short time frame, amplifying a targeted sequence hundreds of millions to billions of times.

polymorphism (see RFLP): The quality or character occurring in more than one form.

posttranscriptional regulation: Any event that occurs after transcription and that influences any of the subsequent steps involved in the ultimate expression of that gene. Reference to posttranscriptional regulation usually refers to events between the termination of transcription and just prior to assembly of the translation apparatus.

posttranslational regulation: Any event that occurs after synthesis of the primary peptide that influences any of the subsequent steps involved in the ultimate expression of the gene. Reference to posttranslational regulation usually refers to the efficiency of the events that modify a peptide, including, but not limited to, glycosylation, methylation, and hydroxylation.

precursor RNA: An unspliced RNA molecule; the primary product of transcription.

primer: A short nucleic acid molecule that, upon base pairing with a complementary sequence, provides a free 3′-OH for any of a variety of primer extension-dependent reactions.

probe: Usually, labeled nucleic acid molecules, either DNA or RNA, that are used to hybridize to complementary sequences in a library, or that are among the complexity of different target sequences present in a nucleic acid sample, as in the Northern analysis, Southern analysis, or nuclease protection analysis. In forensics, a short segment of DNA is used to detect certain alleles. The probe hybridizes, or matches up to, a specific complementary sequence, allowing for the visualization of the DNA complex by either a radioactive "tag" (RFLP) or biochemical tag (HLA DQA1). A single-locus probe marks a specific site (locus), whereas a multilocus probe marks multiple sites.

prokaryote: A microorganism (cell) that has no nucleus (the DNA is not enclosed within a membrane) and lacks other organelles found in the cells of higher organisms.

promoter: A DNA sequence associated with a particular locus at which RNA polymerase binds at the onset of transcription. Promoters typically consist of several regulatory elements involved in the initiation, regulation, and efficiency of transcription.

protein: A molecule composed of amino acids arranged in a specific order determined by the genetic code.

radiochemical: A chemical containing one or more radioactive atoms.

recessive: A gene or allele that is "masked" by another, dominant allele.

recombinant DNA: The hybrid DNA produced in the laboratory by joining pieces of DNA, frequently from different organisms.

renaturation: The reassociation of denatured, complementary strands of DNA or RNA.

replication: The formation of an exact copy. DNA replication occurs when each strand acts as a template for a new, complementary strand, formed according to base-pairing rules.

restriction endonuclease: A class of enzymes that recognizes a specific base sequence (usually four to six base pairs in length) in a double-stranded DNA molecule and cuts both strands of the DNA at every site where this sequence occurs.

restriction endonuclease recognition site: The site where a specific restriction endonuclease cuts the DNA molecule.

restriction fragment length polymorphism (RFLP): The presence of variants in the size of DNA fragments produced upon restriction enzyme digestion due to a change in bases. These different-sized fragments may result from an inherited variation in the distribution of restriction endonuclease sites. RFLPs are used in the laboratory for human identification or parentage determination.

reverse transcriptase (RT): A class of enzymes that catalyze the formation of DNA strands from RNA templates. The name "reverse transcriptase" is from its ability to "reverse" the normal first step of gene expression, that is, the formation of an RNA strand from a DNA template. Once the first DNA strand has been synthesized, it serves as the template for the enzymatic synthesis of the second complementary DNA strand.

RFLP analysis: A technique that uses single-locus or multilocus probes to detect variation in a DNA sequence according to differences in the length of segments created by cutting DNA with a restriction enzyme.

ribonuclease (RNase): A family of resilient enzymes that rapidly degrade RNA molecules. Control of RNase activity is a key consideration in all manipulations involving RNA.

ribonucleic acid (RNA): A chemical found in the nucleus and cytoplasm of cells. A polymer of ribonucleoside monophosphates, synthesized by an RNA polymerase. RNA is the product of transcription and plays an important role in protein synthesis.

ribosomal RNA (rRNA): The predominant class of RNA in the cell. The highly abundant nature of rRNA makes it a useful indicator of sample integrity, quality, and probable utility. The low complexity of this RNA species also makes it useful as a molecular weight marker for RNA electrophoresis.

ribozyme: An RNA molecule with the capacity to act as an enzyme.

RNA polymerase: An enzyme responsible for the synthesis of RNA polynucleotides by the process of transcription, using DNA as a template.

saline sodium phosphate-EDTA (SSPE): A salt solution frequently used for blotting of nucleic acids. It is also an essential component of various hybridization buffers and posthybridization filter washes. The phosphate in this buffer mimics the phosphodiester backbone of nucleic acids, thereby providing enhanced blocking, lower background, and higher signal-to-noise ratio on membranes during Northern and Southern analyses.

sequence: The order of nucleotides (A, C, G, and T) in a nucleic acid or DNA molecule.

single nucleotide polymorphism (SNP): A change in the DNA in which a single base or nucleotide differs from the usual base at that position.

slot blot analysis: A membrane-based technique for the quantitation of specific RNA or DNA sequences in a sample. The sample is usually slot-configured onto a filter by vacuum filtration through a manifold. Slot blots lack the qualitative component associated with electrophoretic assays. *See also* **dot blot analysis**.

sodium dodecyl sulfate (SDS): An ionic detergent commonly used to disrupt biological membranes and to inhibit RNase.

Southern analysis: A technique for transferring electrophoretically chromatographed DNA from an agarose gel matrix onto a filter paper for subsequent immobilization and hybridization. The information gained from Southern analysis is used to qualitatively and quantitatively assess the organization of specific genes or other loci.

specific activity: The amount of radioactivity per unit mass of a radioactive material. It is most frequently expressed in curies per millimole of material (Ci/mmol).

splicing: Ligating two fragments of DNA or RNA end to end to create a new molecule.

Stoffel fragment (AmpliTaq DNA polymerase): A thermostable recombinant DNA polymerase that is smaller (by 289 amino acids) than the full-length AmpliTaq polymerase. The Stoffel fragment of AmpliTaq is more thermostable, has activity over a broader range of Mg++ concentrations, and lacks a 5′ to 3′ exonuclease activity. It is commonly used in multiplex PCR applications.

stringency: A measure of the likelihood that double-stranded nucleic acid molecules will dissociate into their constituent single strands; it is also a measure of the ability of single-stranded nucleic acid molecules to discriminate between other molecules that have a high degree of complementarity and those that have a low degree of complementarity. High-stringency conditions favor stable hybridization only between nucleic acid molecules with a high degree of complementarity. As stringency is lowered, a proportional increase in nonspecific hybridization is favored.

SYBR Green: One member of a new family of dyes for staining nucleic acids. Commonly prepared as a 10,000X stock solution in DMSO, SYBR Green is diluted to a working concentration of 1X in Tris buffer, such as 1X TAE. Among the advantages of using SYBR Green are greatly reduced background fluorescence, higher sensitivity, and reduced mutagenicity when compared with ethidium bromide. SYBR Green I is used to stain DNA, whereas SYBR Green II is used to stain RNA.

Taq DNA polymerase: Thermostable DNA polymerase from the organism *Thermus aquaticus*. Taq is one of several enzymes that can be used to support the polymerase chain reaction.

target: Single-stranded DNA or RNA sequences that are complementary to a nucleic acid probe. Target sequences may be immobilized on a solid support or may be available for hybridization in solution.

template: A macromolecular informational blueprint for the synthesis of another macromolecule. All polymerization reactions, including replication, transcription, and PCR, require templates; these dictate the precise order of nucleotides in the nascent strand. Primer extension-type reactions cannot proceed in the absence of template material.

termination codon: Codons (e.g., UAG, UGA) that signal termination of the synthesis of a polypeptide chain (UAG, UGA).

T_m: Melting temperature; that temperature at which 50% of all possible duplexes are dissociated into their constituent single strands. To facilitate formation of all possible duplexes, hybridization is conducted below the T_m of the duplex; the lower the temperature, the greater the likelihood that duplexes, including those with mismatches, will form.

transcription: The transfer of information from various parts of the DNA molecule to new strands of rRNA, tRNA, or mRNA, which then carry this information from the nucleus (in eukaryotes) into the cytoplasm.

transduction: Transfer of genetic material from one cell to another by means of a viral vector.

transfer RNA (tRNA): A moderately abundant class of RNA molecules that shuttle amino acids to the aminocyl site of the ribosome during protein synthesis. The total mass of tRNA in the cell is occasionally assayed as a housekeeping indicator of transcription, in order to show that a particular experimental manipulation has not resulted in a change in overall transcription in the cell.

transformation: Introduction of exogenous DNA into a cell, causing it to acquire a new phenotype, as in bacterial transformation.

translation: The process by which peptides are synthesized from the instructions encoded within an RNA template. Translation occurs as mRNA is deciphered by the ribosomes.

triploid: Having three complete sets of chromosomes (three of each chromosome). *Compare to* **diploid** *and* **haploid**.

tRNA: *See* transfer RNA.

upstream: Sequences in the 5′ direction (away from the direction of expression) from some reference point. For example, the 5′ cap in eukaryotic mRNA is located *upstream* from the initiation codon.

UV light (ultraviolet light): Short-wave, high-energy portion of the electromagnetic spectrum. Because nucleic acids absorb light maximally in the ultraviolet range (260 nm), samples of nucleic acids are stained with dyes (e.g., EtBr) and irradiated with UV light for visualization. **CAUTION:** UV light is mutagenic and can severely damage the skin and the retina of the eye. Be certain to wear proper eye and skin protection at all times.

variable number of tandem repeats (VNTR): Multiple copies of virtually identical base pair sequences, arranged in succession at a specific locus on a chromosome. The number of repeats varies from individual to individual, thus providing a basis for individual recognition.

vector: A nucleic acid molecule such as a plasmid, bacteriophage, or phagemid into which another nucleic acid molecule (the so-called insert or foreign DNA) has been ligated. Vectors contain sequences that, in a suitable host, permit propagation of the vector and the DNA that it carries.

Western analysis: A technique for transferring electrophoretically chromatographed protein from a polyacrylamide gel matrix onto a filter paper for subsequent characterization by antigen–antibody recognition. The information gained from Western analysis is used to qualitatively and quantitatively assess the prevalence of specific polypeptides.

References and Suggested Reading

Anderson, S., A. T. Bankier, B. G. Barrell, M. H. de Bruijn, A. R. Coulson, J. Drouin, I. C. Eperon, D. P. Nierlich, B. A. Roe, F. Sanger, P. H. Schreier, A. J. Smith, R. Staden, and I. G. Young. 1981. Sequence and organization of the human mitochondrial genome. Nature 290:457-465.

Andrews, R. M., I. Kubacka, P. F. Chinnery, R. N. Lightowlers, D. M. Turnbull, and N. Howell. 1999. Reanalysis and revision of the Cambridge reference sequence for human mitochondrial DNA [letter]. Nat. Genet. 23:147.

Butler, J. M., and B. C. Levin. 1998. Forensic applications of mitochondrial DNA. Trends in Biotech. 16:158-162.

Edwards, A., A. Civitello, H. A. Hammond, and C. T. Caskey. 1991. DNA typing and genetic mapping with trimeric and tetrameric tandem repeats. Am. J. Hum. Genet. 49:746-756.

Edwards, A., H. A. Hammond, L. Jin, C. T. Caskey, and R. Chakraborty. 1992. Genetic variation at five trimeric and tetrameric tandem repeat loci in four human population groups. Genomics 12:242-253.

Gilbert, D. A., Y. A. Reid, M. H. Gail, D. Pee, C. White, R. J. Hay, and S. J. O'Brien. 1990. Application of DNA fingerprint for cell-line individualization. Am. J. Hum. Genet. 47:499-514.

Gill, P., A. J. Jeffreys, and D. J. Werrett. 1985. Forensic application of DNA "fingerprints." Nature 318:577-579.

Jeffreys, A. J., V. Wilson, R. Neumann, and J. Keyte. 1988. Amplification of human minisatellites by polymerase chain reaction: Towards DNA fingerprint of single cells. Nucleic Acid Res. 16:10953-10971.

Jeffreys, A. J., V. Wilson, and S. L. Thein. 1985. Individual-specific fingerprint of human DNA. Nature 316:76-79.

Kasai, K., Y. Nakamura, and R. White. 1990. Amplification of a variable number of tandem repeat (VNTR) locus (pMCT118) by the polymerase chain reaction (PCR) and its application to forensic science. J. Forensic Sci. 35:1196-1200.

Miller, S. A., D. D. Dykes, and H. F. D. Polesky. 1988. A simple salting out procedure for extracting DNAs from human nucleated cells. Nucleic Acid Res. 16:1215.

Nakamura, Y., M. Carlston, V. Krapcho, and R. White. 1988. Isolation and mapping of a polymorphic DNA sequence (pMCT118) on chromosome 1p (D1S80). Nucleic Acid Res. 16:9364.

Nakamura, Y., S. Gillilan, P. O'Connell, M. Leppert, G. M. Lathrop, J-M. Lalouel, and R. White. 1987b. Isolating and mapping of a polymorphic DNA sequence PYNH24 on chromosome 2 (D2S44). Nucleic Acid. Res. 15:10073.

Nakamura, Y., M. Leppert, P. O'Connell, R. Wolfe, T. Holm, M. Culver, C. Martin, E. Fujimoto, M. Hoff, E. Kumlin, and R. White. 1987a. Variable number of tandem repeat (VNTR) markers for human gene mapping. Science 235:1616-1622.

Sanger, F. S., S. Nilken, and A. R. Coulson. 1977. DNA sequencing with chain-terminating inhibitors. PNAS 74:5463-5467.

Southern, E. 1975. Detection of specific sequences among DNA fragments separated by gel electrophoresis. J. Mol. Biol. 98:503–527.

Wahls, W. P., L. J. Wallace, and P. D. Moore. 1990. Hypervariable minisatellite DNA is hotspot for homologous recombination in human cells. Cell 60:95-103.

Warne, D., C. Watkins, P. Bodfish, K. Nyberg, and N. K. Spurr. 1991. Tetranucleotide repeat polymorphism at the human β-actin related psuedogene 2 (actbp2) detected using the polymerase chain reaction. Nucl. Acids Res. 19:6980.

Index

NATIONAL CURRICULUM
Gymnastics

Anne
Williams

Hodder & Stoughton

A MEMBER OF THE HODDER HEADLINE GROUP

My thanks to the staff and pupils of Ninestiles School who helped so patiently with the photographs.

British Library Cataloguing in Publication Data

Williams, Anne, 1947–
   National Curriculum gymnastics
   1. Gymnastics – Study and teaching (Secondary) – Great
Britain
   I. Title
   796.4'4'0712'41

ISBN 0 340 67377 X

First published 1997
Impression number 10 9 8 7 6 5 4 3 2 1
Year   2000 1999 1998 1997

Typeset by Fakenham Photosetting Limited, Fakenham, Norfolk
Printed in Great Britain for Hodder & Stoughton Educational, a division of Hodder Headline Plc, 338 Euston Road, London NW1 3BH by The Bath Press, Bath, Somerset

Contents

Why gymnastics?

Introduction

Gymnastics has been a part of the physical education curriculum through a period which has seen major changes in philosophies of education and in the organisation of schools. It is identified in national curriculum documentation as one of the core areas of the physical education curriculum, along with games, dance, swimming, athletics and outdoor and adventurous activities. These activities constitute the means whereby the aims of physical education are to be achieved. It will be useful to clarify the aims of physical education in order to identify the particular contribution which gymnastics can make. The working group set up to advise on the national curriculum for physical education (DES 1990) provides a summary of physical education's contribution to the whole curriculum when it states that physical education:

★ develops physical competence;

★ promotes physical development;

★ teaches pupils, through experience, to know about and value the benefits of exercise;

★ establishes self-esteem through the development of physical confidence;

★ develops artistic and aesthetic understanding within and through movement;

★ helps pupils to cope with both success and failure in the context of co-operative and competitive physical activities;

★ provides experience of physical activities which should lead pupils to lifelong participation.

The group considers that physical education also contributes to:

★ the development of problem-solving skills;

★ the development of inter-personal skills;

★ the forging of links between the school and the community and across cultures. (DES 1990, p. 5)

Aims and purposes of gymnastics

This section considers the ways in which gymnastics may contribute to the achievement of the aims of education and of physical education. A number of possibilities are discussed, and, in some schools, given resources and commitment, it may be that all of these possibilities can be explored. In other schools, for various reasons, the curriculum time available for gymnastics may be limited, and, consequently, the number of avenues which can reasonably be explored will be restricted. In these circumstances, the aims and features which take precedence will depend upon the philosophy of the school and the physical education programme and upon the priorities adopted by the teacher. Clearly what can be achieved in a school which opts to offer gymnastics as a half unit at Key stage 3 with no opportunity to continue the activity at Key stage 4 will not be the same as the school which chooses to provide a full unit of study at key stage 3 followed by gymnastics both as a GCSE activity and in a non-examination programme at Key stage 4.

A number of the features claimed for gymnastics are also, of course, rightly claimed for other activities which may form a part of the physical education curriculum. It is therefore important to examine the purposes of gymnastics in a specific school, in relation to the rest of the physical education curriculum and, indeed, in relation to the whole curriculum of that school. For example, it is possible to teach gymnastics in such a way that its aesthetic qualities receive great emphasis. This might be wholly appropriate in a school which, at Key stage 3, has chosen to offer gymnastics and not dance, and where other opportunities for aesthetic activity are limited. Another school, however, may have chosen to offer both dance and gymnastics at Key stage 3 and may have a thriving performing arts faculty catering for aesthetic experience through a variety of media. It could well be that other features of gymnastics might be given a higher priority in the latter school.

Gymnastics focuses on the body. In so doing it differs from games which are concerned with manipulating other tools (rackets, sticks, balls, etc.) and which require the efficient use of the body in order to achieve the purposes of the game. Gymnastics aims specifically to develop body management and body awareness. It is concerned with bodily skill and with precision of movement and form. Work with young children concentrates on body management in the sense of discovering possibilities and also becoming aware of capabilities and limitations. As pupils gain experience, demands in terms of control, finish and precision of movement can be increased. It is this concern with precision of movement and with quality of response that distinguishes gymnastics from other activities such as keep fit, aerobics, circuit training and so on, which also focus on the body but which have other main purposes.

The control, style and precision which characterise good quality gymnastics can be seen readily in the performance of competitive gymnastics. Curriculum gymnastics can make exactly the same demands in terms of quality of response, although the level of difficulty will obviously be significantly different. Indeed it may be argued that unless high quality movement is demanded, then other activities might equally well fulfil many of the purposes which may be attained through the medium of gymnastics.

Much criticism levelled at contemporary physical education programmes is of the undue emphasis placed upon competition. Gymnastics' place in the curriculum may be argued on the grounds that it is one way of providing a balance between competitive and non-competitive activity. While the judicious use of competition can be a useful educational tool and an important motivating force, intrinsically competitive activities, particularly team activities, when played to adult codes of rules, are disliked by significant numbers of pupils, particularly the less successful. Gymnastics is one area of the curriculum which offers scope for non-competitive activity, assessed according to individual improvement and progress. Quality of performance may be assessed without the need to produce winners and losers. If the physical education curriculum as a whole

Gymnastics offers scope for co-operative activity

is to provide an experience relevant to and appropriate for all pupils, then this kind of activity balance is important.

Gymnastics also contributes to a balanced curriculum in that it offers scope for both individual and shared or co-operative activity. This is in contrast to team activities which continue to dominate physical education and which are likely to increase in overall emphasis as a result of current government initiatives (Department of National Heritage 1995). The individual nature of gymnastics means that it is possible to offer children the opportunity to progress at their own pace. The wealth of material available to the gymnastics teacher is such that a selection can be made to suit all ages, abilities and aptitudes. This wide variety of material derives from many sources, from the work of Ling and Jahn, from that based upon Laban's principles of movement, from earlier curriculum work and from the world of competitive gymnastics.

The opportunity to co-operate and share is present in many physical activities including gymnastics, where the kind of co-operation demanded is rather different from that required in, for example, the team game. Pupils have to share the available work space and also the available equipment and have to co-operate with one another in setting out or rearranging the work space. This aspect of co-operative behaviour is present in all gymnastics work. Co-operation may also be required if partner or group work features in the gymnastics programme, and the ability to share ideas, to assist one another and to respect the views and ideas of others may be developed through this medium.

Gymnastics can maintain or improve strength, endurance and flexibility as well as offering opportunities for using these three physical qualities in the learning and performance of different activities. It also provides an excellent medium for teaching about the effects of different kinds of activity upon the body. In the context of offering a curriculum which is relevant to all pupils, the variety of activity available within the gymnastics context means that success can be readily achieved by those with vastly different physiques and physical capacities.

Gymnastics in an educational context is, of course, concerned with more than the performance of a number of specific skills and agilities, important though this element of the programme is. The process of learning and the teaching methods used are of equal significance. One of the features of contemporary curriculum gymnastics is the variety of teaching styles and strategies which may be utilised. The first National Curriculum Order acknowledged this more or less explicitly by use of language such as 'be enabled to, be guided towards, be given opportunity to' (DES 1992). The current Order (DFE 1995) simply states that pupils should 'be taught'. However, it should be remembered that in order to maximise the performance of all pupils and to engage them in the process of planning, performing and evaluating, as required by the current National Curriculum Order, a range of teaching styles and strategies will be needed.

The process model for physical education applied to gymnastics

The process model is a cyclical model as illustrated in Figure 1. It underpins National Curriculum physical education as stated in the general requirements for all key stages:

Opportunities for using strength

Opportunities for using flexibility

Physical education should involve pupils in the continuous process of planning, performing and evaluating. This applies to all areas of activity. (DFE 1995, p. 22)

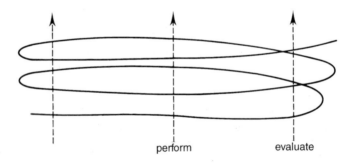

perform   evaluate

Fig 1 The process model

Jane's gymnastics experience at Key stage 3 (see the boxed study on the next page) illustrates this in practice.

Although they are inextricably linked in practice, it may be helpful to consider each component of the process separately.

Planning

In the context of the National Curriculum Order, planning may be for individual actions, for sequence work or for partner or group work.

Plan – Perform – Evaluate

Jane arrives at secondary school with a limited range of securely learned gymnastics skills and the ability to observe and describe perceptively. She is able to plan her work in order to share space and apparatus effectively and to compose simple sequences. By the end of Year 7 Jane is able to plan an apparatus sequence which uses box, two benches and four mats so that she can perform her piece of work at the same time as two other pupils and so that certain prescribed activities are included. She is able to include previously learned skills using mats and benches and new skills on the box: rolling backwards off the end of it and a handstand forward roll from lying across the top. She is able to watch the performance of others and to describe it accurately in terms of the skills and actions included. She is also able to make judgements about the quality of the performance of the individual activities within the sequence and about the extent to which the sequence shows continuity.

By the end of Year 9 Jane is able to contribute to the planning of a group sequence which includes partner balances, more advanced apparatus skills and a combination of canon and unison work. Discussion covers who should work with whom for a variety of partner balances which the group wants to include and what apparatus skills different group members can include to give variety within the sequence. Jane is able to perform a number of partner balances as the 'balancer' and can act as 'support' for one other member of the group who is physically smaller than her. Using a trampette she can now perform a handstand forward roll along the box. With her group she draws up criteria by which she will 'assess' the work of other groups. After watching others' sequences, the group discusses them in relation to their criteria and presents a brief report to the rest of the class.

For individual movements

Planning in many contexts is, effectively, 'mental rehearsal' and as such is important for successful performance. The planning element may be important to the successful completion of an activity, to more effective performance (e.g. better quality, greater height or distance), to safe performance (e.g. safe landings) or to enabling one action to prepare for a subsequent one. For example, in the process of learning or refining performance the pupils will need to plan to position the feet close to the hips in order to stand up at the end of a forward roll. They will need to plan to take off from two feet rather than one foot in order to use a springboard effectively. They will need to plan to let the hips go beyond the vertical before flicking when performing a headspring. Once the skill is consolidated this conscious planning process will no longer be needed.

Planning for sequence work

Sequence work involves the selection of actions to be included, decisions about ordering for maximum quality and interest, decisions about timing and pathways, particularly where the working space is being shared and several pupils need to work at the same time. All of these involve planning. For example, planning the order in which activities are to be performed in a sequence will lead to good continuity and quality.

Planning for pair and group work

Work in pairs or larger groups involves additional considerations and makes planning more explicit because decisions have to be shared. For example, actions to be included in a matching pair sequence have to be planned with regard for the capabilities and limitations of both partners. In a large group sequence planning involves deciding who will take what role, for example, if partner balances are to be included who will be the bases and who will be the balancers.

Performing

Performing involves the skills, knowledge and understanding needed to undertake a range of gymnastics activities successfully:

★   it includes performance of single actions or skills – forward roll, front somersault;

★   it includes the performance of these skills in a sequence in gymnastics or dance and the ability to perform increasingly complex sequences. These can also involve work with a partner or in a group;

★   it involves the ability to conform to rules or conventions. In the context of gymnastics this means performing generic skills according to generally agreed conventions, e.g. handstand with straight body position when the pupil has sufficient skill to achieve this;

★   it involves the ability to sustain performance over a period of time – practising a skill;

★   it involves being able to refine skills through practice or rehearsal;

★   it involves the ability to adapt to new situations, e.g. performing a cartwheel at a different speed from usual in order to synchronise with a partner, performing a forward roll along a high platform;

★   it involves understanding the principles of effective performance in order to sustain a high level of expertise, e.g. knowing that the body turns more easily in a streamlined position, therefore a turning jump should be attempted with legs together and arms close to the body.

Performance develops in a cyclical way:

★ establish;

★ adapt;

★ refine;

★ vary.

At Key stages 3 and 4, the process may begin at refinement, seeking increased quality of performance in established skills, or variations to apply them in a new context, for example, headstand or handstand into forward roll performed along a box where previously this had been performed on the floor; backward roll performed in unison as part of a group sequence where previously it had been performed as a solo skill. Where time and teaching expertise at Key stage 2 has been limited, adaptations and variations of quite basic gymnastics actions will need to be covered if pupils are to develop the confidence and competence to progress successfully to more advanced work. Much work at Key stage 2 will involve encouraging pupils to extend their personal movement repertoire within the activity categories specified in the programme of study.

Later work may still involve the establishment and adaptation phases, for example, learning skills made possible by the availability of new apparatus, e.g. climbing ropes for those whose primary school did not offer this opportunity; use of reuter boards and/or trampettes; learning more advanced skills such as handspring, headspring, flic flac.

Evaluating

Evaluating involves making judgements about one's own performance or that of others which will help in future planning and performance:

★ it involves making simple judgements about performance against given criteria – were his feet together? Did she take off from one or two feet?

★ it involves recognising others' success – how many balances did your partner manage to do? How long did your partner hold that balance for?

★ it involves making judgements about performances against given criteria – were there three balances in that sequence, and what linking actions were used? How successful was the group in synchronising their actions?

★ it involves making judgements about safe practice – is that apparatus too close to the wall? Have I got room to do a cartwheel here without kicking anyone?

★ it involves making comparisons about performance – who got most height off the trampette? Why do you think that was?

Evaluation involves observation and analysis

★   it involves monitoring and adjusting one's own performance – if I stand up on the other foot I will be in a better position to do the cartwheel;

★   it involves analysing performance against criteria which are at least partly identified by the pupil;

★   it involves identifying the key aspects of a performance which need further practice.

Evaluation skills should be developed systematically:

Observe
'Watch this demonstration.'
'Watch your partner while you are waiting for your turn.'

Describe
'Watch this demonstration and tell me what Jill's legs are doing.'
'Watch this demonstration and tell me what shape David's body makes.'
'Watch this demonstration and tell me what actions Sarah and John have used.'

Analyse
'Watch David and tell me where his hips are when he pushes away from the box.'
'Watch Melanie and tell me where her hips are in relation to her shoulders.'
'Watch this group of three and tell me what they have done so that they can keep their actions synchronised.'
'Watch your partner and work out why they are finding standing up difficult.'

Compare
'Watch these two handstands. Why can Sarah keep her balance longer than Anne?'
'Watch these two sequences. Why do you think that I would give Malcolm more marks than Jason?'

Make judgements using given criteria
'On the chalkboard are some points to help you to improve your handstands. Watch your partner and use these points to help them.'
'Your task was to produce a sequence using every piece of your apparatus and including balancing both the right way up and upside down, taking your weight on your hands, and jumping actions. Watch this group and tell me whether they have succeeded.'

Make judgements using self-chosen criteria
'With your partner, decide how you would assess the work that you have just been doing, then watch each other and apply the criteria you have chosen.'
'You are going to watch each other's group sequences and give a report on how you think that they have done. Spend the next few minutes discussing what criteria you are going to use to assess them.'

By Key stage 3 pupils should be capable of observing, describing and analysing, although those who have not had this opportunity in the gymnastics context at Key stage 2 will need help. Key stage 2 pupils should have had experience of making simple comparisons and making judgements against given criteria but this should not be assumed.

The nature of curriculum gymnastics

Unlike some other physical activities, there are significant differences between much of the work undertaken in curriculum gymnastics and that presented to the pupil as the adult world of competitive gymnastics work. This section will consider the similarities and differences between various gymnastics forms.

Through the 1950s and 1960s, gymnastics in the curriculum developed independently of competitive gymnastics, particularly in girls' physical education, where 'educational gymnastics', based on Laban's principles of movement, had gained much ground. Boys' physical education at this time had either retained traditional vaulting and agility work or adopted the problem-solving approach advocated by Bilbrough and Jones. Following the explosion of interest which occurred in the wake of the Munich Olympic Games of 1972, various aspects of competitive gymnastics previously excluded from most curriculum work began to appear in schools. Opportunities to take part in gymnastics outside school through clubs increased significantly, especially for girls. For some time, developments in gymnastics as a competitive medium continued apace, while curriculum gymnastics showed much less evidence of change.

It was perhaps unfortunate that conflict between supporters of different gymnastics forms led to a focus on disparities between them, where examination of common ground might not only prove more fruitful but would also reveal that certain assumptions about the different forms are mistaken or at least open to misinterpretation. Because gymnastics encompasses a wide range of material and because education involves many different kinds of experiences, there are numerous ways in which gymnastics can be taught as an educational medium. It is therefore difficult to identify one particular approach which is 'best' for curriculum gymnastics. Furthermore, staff and resources vary and pupils enter school with widely differing abilities, interests and previous experience.

There is little doubt that pupils are interested in gymnastics and equally little doubt that this interest can quickly be lost if programmes offered in school fail to capitalise on this interest. Television coverage reveals ever-rising standards of performance and the diversification of what was previously called Olympic Gymnastics into various competitive forms: Artistic Gymnastics, Modern Rhythmic Gymnastics, Sports Acrobatics. It is inevitable that the way pupils perceive gymnastics will be coloured by media images as well as by the experiences offered in school. The result may be extreme disappointment that the school programme cannot offer the challenge or excitement generated by television performances, or relief that gymnastics consists of more than the technically demanding skills of the international performer.

Gymnastics is however faced with a dilemma in that the obvious relation between the curriculum activity and the 'adult' activity form does not exist. This has advantages and disadvantages for the teacher. Trying to teach a modified game to pupils who are not ready to play the adult version but who have an image of the adult professional game against which anything else is a poor second best can be a challenge to the best of teachers. Nevertheless this link does mean that the activity has a relevance which is not as immediately obvious, where the activity appears to have little in common with that practised under a similar label in the outside or adult world. It should be noted that it is largely male activities which enjoy these close links and that for most female pupils, links between their curriculum experience and out-of-school interests tend to be much more tenuous.

There is much which could and should be common both to competitive forms of gymnastics and to curriculum work, together with a number of differences which are of degree rather than of kind. Both forms of gymnastics should be concerned with the development of skilled body management. Unfortunately, much of the criticism which is levelled at curriculum work is of its failure to develop this aspect of work. The 'pick a number of basic movements and join them together to form as many sequences as possible' approach is criticised by Groves (1973) who also points out that this approach all too often leads to a situation where pupils lose enthusiasm because they lack challenge. Conversely, competitive gymnastics is often claimed to lack relevance in the school situation because of its high skill threshold.

All too often there appears to be a failure to make a sufficiently clear distinction between 'quality' and 'difficulty'. No one would challenge the fact that many of the skills of competitive gymnastics are extremely demanding technically or that they have a very high skill threshold. Observation of many school lessons, however, suggests that there is often a failure to appreciate that high quality can be achieved where the level of difficulty is low, just as much as in actions which make great technical demands. The achievement of skilled body management implies high quality performance and should be of just as much concern to the school gymnastics teacher as to the competitive gymnastics coach. The levels of difficulty demanded may well differ.

The competitive gymnast needs to acquire skills of preordained levels of difficulty in order to meet the requirements laid down by the sport's governing body. The teacher is not bound by such constraints. In any case, the size of most classes means that concentration on actions requiring a 1:1 teaching situation is simply not an effective use of the teacher's time and expertise. The freedom which the teacher has to select from our rich gymnastics heritage, notwithstanding the prescription of the National Curriculum Order, is potentially one of the great strengths of curriculum gymnastics.

Many past publications on gymnastics have concentrated chiefly on the material to be used, much of which derives from Laban's movement principles, developed into a thematic approach. This emphasis on content has, unfortunately, tended to divert attention from the real key to curriculum gymnastics, namely the process of learning or the style of teaching. While the National Curriculum emphasises performance, it also includes the requirement to consider both planning and evaluation throughout all programmes of study. If pupils are to become independent learners, capable of planning and evaluating their own performance, then the process by which learning takes place is crucial.

It is this which is the major difference between curriculum work and competitive gymnastics which, by its nature, focuses more upon the end product than on the process by which it is achieved. This is evident in the ways in which achievement is assessed. While award schemes, which are criterion-referenced, can be achieved by any number of pupils, the skills nevertheless have to be prescribed. In competitive work, the performer with the highest mark is deemed to be the best. Curriculum work can be assessed in much broader performance terms and can include the assessment of understanding, of planning and evaluating, and of broader factors such as personal and social skills.

As far as the material of gymnastics is concerned, the distinction between curriculum work and competitive gymnastics need not be a difference in kind. The real difference lies in the freedom from governing body constraints enjoyed by the teacher. Evaluation of a pupil's achievement can be in relation to his or her ability and based upon a more flexible range of possible activities rather than according to preordained conventions in terms of technical requirements or style. Additionally, this freedom means that the teacher can build on the pupil's strengths and avoid focusing to excess on weaknesses. For example, choices towards the end of Key stage 3 can enable one pupil to work on activities which exploit his or her upper body strength while another may prefer to refine performance in activities requiring flexibility. What is important is that this freedom is used to exploit abilities and not as an excuse for mediocrity. It is equally important that technical requirements needed for progression to more advanced work are understood so that freedom of choice does not militate against future progress.

Ultimately, gymnastics in the curriculum is one means whereby some of the aims of physical education may be achieved. These aims in turn reflect those of the total educational process. The National Curriculum is based upon the principle that all pupils are entitled to a broad, balanced, relevant and differentiated curriculum.

CHAPTER 2

The material of gymnastics

Before detailed consideration of curriculum content can begin, an understanding of the material of gymnastics is needed, so that any selection of content may be made with an awareness of how what is chosen relates to the whole field of gymnastics activity. Just as discussion of the nature of curriculum gymnastics is hampered by focusing on disparities rather than by seeking similarities and common ground, so understanding of the material of gymnastics may be hindered if classifications or isolated approaches do not fit into a particular approach.

Curriculum gymnastics in England has used classifications of material based upon anatomical considerations, upon effects of activities upon the body and upon Laban's movement principles.

Anatomical considerations formed the basis of many early classifications of gymnastics material. Curriculum work in England during the early part of this century was based upon the Swedish system developed by P.H. Ling, whose influence can be seen in PT syllabuses of the time. The 1909 Syllabus, for example, divides exercises into:

1 introductory exercises;

2 trunk bending backward and forward;

3 arm bending and stretching;

4 balance exercises;

5 shoulder blade exercises;

6 abdominal exercises;

7 trunk turning and bending sideways;

8 marching, running, jumping, games, etc.;

9 breathing exercises.

The system of exercising each part of the body in turn was retained in the 1933 Syllabus. While *Moving and Growing* and *Planning the Programme* (Ministry of

Education 1952 and 1953) marked a significant departure from such rigidly laid down guidelines, a framework was still offered (to those teachers who desired one) in which anatomical considerations were retained:

1 general activity;

2 compensatory movements:
   a) trunk movements;
   b) arm and shoulder girdle movements;
   c) foot and leg movements;

3 agility movements.

Dissatisfaction in some quarters with the limitations of purely anatomical considerations led to classifications of activities based upon the effects of activities on the body. In the 1930s both Knudsen and Bukh divided activities according to their effect on the development of strength, suppleness and ability. Endurance or stamina has also featured in this kind of classification. It is probably true to say that these aspects of gymnastics were of more interest to male physical educationists than to females. Nevertheless, *Planning the Programme* acknowledges the relevance of this kind of classification for many teachers when noting that a framework based upon terms such as mobility, strength and agility has often been found useful.

Both of these classifications see gymnastics as a system of exercises. Development of gymnastics as a sport was quite separate, stemming from Jahn's work in Germany and virtually disappearing from England as Ling's system was adopted by the Ministry of Education. Munrow's comments made in 1955 serve to underline the extent of the separation.

In England the word gymnastics has become inseparably connected with the system and the sport has declined almost to the point where a name is no longer necessary for it. Brave enthusiasts who urge its revival – may success crown their efforts – now give it the name Olympic Gymnastics.

It is perhaps small wonder that skills associated with sports gymnastics received so little attention when Laban's work began to have a major influence on curriculum work in the 1950s and 1960s. Laban's classification of movement was intended to be all-embracing, although his own writing was with reference to dance and to industry. Many educational gymnastics texts use Laban's work either explicitly or implicitly. One of the earliest to put pen to paper was Ruth Morison who went on to publish a more comprehensive text in 1969. After pointing out that the material of educational gymnastics was drawn from the natural everyday activities of children, she discusses three factors to which she gives equal importance:

1 bodily aspects of action;

2 dynamic aspects of action;

3 spatial aspects of action.

To these may be added relationships, that is, partner and group work. The material therefore corresponds to Laban's classification of dance work into Body, Effort, Space and Relationship. From this very general classification, material is organised thematically. Unfortunately, themes suggested by writers not only vary, but also, at times, contain some inconsistencies. It is difficult for the inexperienced to understand how the same action can appear as 'locomotion' in one classification, as 'weight transference' in another and as 'turning' in a third. More serious perhaps are divisions such as twisting and turning where actions which involve twisting when examined biomechanically (for example a full-turn jump) are taught as part of work on turning.

Nevertheless the thematic approach, underpinned by Laban's movement principles, dominated women's gymnastics teaching during this period (see for example, Maulden and Layson 1979 and Williams 1973). It is interesting to note, however, that the London County Council's (1965) publication on educational gymnastics, while using a thematic structure and Laban's motion factors, continues to advocate the division of tasks within a lesson into 'whole body work', 'work with weight supported on the arms', and 'leg work', thereby retaining some element of more formal work. More typical of the content analysis of educational gymnastics would be Morison's:

Actions emphasising locomotion:

1 transfer of (rocking, rolling, steplike actions, sliding);

2 travelling (on different body parts, at different speeds, moving and stopping, in different directions and using different pathways);

3 flight (take off, flight, landing, single and double take offs, from other body parts, shapes, landing on one or two feet, on hands, while turning, assisted).

Actions emphasising balance:

1 weightbearing;

2 balancing skills;

3 entrances to and exits from balance;

4 on and off balance.

Control of movement emphasising:

1 bodily aspects through bend/stretch/twist/turn/shapes/symmetry/asymmetry;

2 dynamic aspects;

3 spatial aspects.

Relationships in partner and group work:
Male physical educationists were in the main sceptical of the approach adopted by female members of the profession. The apparent disregard of mobility, strength and endurance was questioned, as was the omission of teaching specific agilities. It is, how-

ever, difficult to find men's work based upon any systematic classification. Much was based on vaulting and agility, using aspects of sports gymnastics (with some adaptations) traceable to quite different roots from that underpinning other work. Where a problem-solving approach was adopted (as in the work of Bilbrough and Jones 1973 and John Cope 1967), lessons tended to retain the sort of structure advocated by the LCC, retaining a balance through ensuring that tasks exercised the whole body fairly systematically.

Recent thinking about the material of gymnastics has been influenced by a number of factors. The growth in participation in sports gymnastics and in television coverage of it has led to far greater awareness of, and knowledge about, the content of competitive work. Physical education teachers are no longer trained in single-sex institutions and consequently the differences in approach between the sexes which were so evident in the 1950 and 1960s have lessened. The time available for gymnastics in higher education institutions has decreased.

A classification of gymnastics material

The classification which follows provides a logical ordering of material which is not restricted to a specific approach or method. It draws heavily upon the work of John Wright (1980). Those anatomical and kinesiological factors which assist in providing a logical ordering of actions are utilised. It also incorporates Laban's classification of movement in order to give due attention to those aspects of space, dynamics and relationships which augment, arise from or modify action of the whole body or of body parts. By drawing on these different features it is possible to offer a classification of material which may be used to describe, analyse or structure gymnastics in any context, and in which any gymnastics movement may be located. In considering curriculum gymnastics, it will be necessary to select from the material which could be utilised, and selection of material is discussed later. Table 2.1 summarises the material which is described in greater detail in the following pages.

Actions

Gymnastics is first and foremost action, in that without the action of the body or of body parts, there can be no spatial, dynamic or relationship variable. Understanding of the actions both of the whole body and of body parts, and of the relationship between the two, is thus fundamental.

Whole body positions

Held positions are used in many gymnastics situations. They may be a starting point or a finishing position. They are frequently adopted, particularly in competition or display work, as demonstrations of strength, mobility, balance or a combination of all three.

They may be considered:

Table 2.1 Gymnastics material classified

Action	Space	Dynamics	Relationships
Whole body positions (i) in different relationships to point(s) of support (ii) Using different body parts as support (iii) stable or unstable	Personal space General space Direction Level Height and distance Pathway	Speed Timing Use of appropriate effort or force	To floor To apparatus To other people
Whole body actions – travelling rolling stepping/step-like springing assisted springing swinging flying dropping sliding part-circling combinations of the above			
Whole body actions – non-travelling circling pivoting step turning spinning swing turning free flight turning lifting lowering tilting/leaning waving swinging rocking springing in place combinations of the above			
Combinations of whole body positions; *whole body actions – travelling;* *whole body actions – non-travelling*			
Actions of body parts bend/stretch open/close circle/partcircle/swing twist shoulder elevation/depression combinations of the above			
Functions of body part actions initiate/maintain/accelerate decelerate/redirect/arrest motion change direction/pathway/plane change relationship of whole body to floor/apparatus/other gymnasts change relationship of gymnast's body parts			

a) in relation to the point or points of support:

★ beneath points of support (for example, hands from high bar, front and rear scale on rings);

★ above points of support (for example, handstand on box, front rest on bar, floor balances);

★ to side of points of support;

★ various combinations from i, ii and iii.

b) in terms of their relative stability.
Positions fall broadly into two categories, namely:

★ relatively *stable;*

★ relatively *unstable.*

Relatively stable positions are commonly called *balances.* The stability of any held position depends upon several factors:

★ the position of the centre of gravity – this must be over the supporting base if the position is to be maintained;

★ the lower the centre of gravity, the more stable the position;

★ the larger the supporting base, the more stable the position.

Thus a large base such as the back is stable and so is a base using four small points of support, such as hands and feet, spread over a wide area. In contrast, a balance on one hand is unstable and difficult to maintain. A headstand on the floor is easier than a headstand on a beam because the three points can be spread more widely on the floor than on a beam.

c) in terms of body parts used:
foot/feet; legs; trunk;
hands; arms; head/neck.

National Curriculum requirements make reference to this aspect of gymnastics in requiring that pupils be taught first different ways of balancing (Key stage 1) and, later, to increase and refine balancing skills (Key stage 3).

Whole body actions – travelling

a) rolling – any action where the body weight is taken in succession by adjacent body parts or by parts which have been made adjacent, such as in placing the feet adjacent to the hips in order to complete a forward roll to stand;

b) stepping and steplike (for example, climbing) – actions where a body part is removed

Travelling through stepping

and replaced (either the same body part as in walking on the hands, or a different body part as in a walkover). All walking actions are including (on feet and on hands); stepping may be from other parts, for example, from knee to knee or including wheeling action such as cartwheel;

c) springing – all actions where the body is launched into the air so that travel occurs in free flight. Thus jumps are included in this category, with various take offs and landing, also springing from feet to hands as in dive rolls, dive cartwheels, or from hands to feet as in many traditional vaults such as headspring, long arm overswing, thief vault and so on. Somersaults and other aerial movements also come into this category;

d) assisted springing – where the spring is assisted, the body does not break contact with the floor or apparatus. There is, however, a sudden increase in force which differentiates this category from stepping or steplike actions. A handstand from a double foot take off onto a box or bench would come into this category;

e) swinging – this uses a pendulum action in order to travel, for example, using a rope to cross from a bench to a beam;

f) flying (from swing or being thrown) – flying as a result of being thrown is seen to great effect in sports acrobatics tempo work where one partner throws the other into

Travelling with flight

somersaults to the floor, or to be recaught. This kind of free flight may also be achieved from a rope or bar by letting go before landing;

g) dropping – a method of travelling down, for example, simply releasing a high bar and dropping to the floor;

h) sliding – sloping apparatus is the most obvious choice for this activity although it can also take place on a suitable floor surface. The part in contact with the floor or apparatus remains in contact and reduction in friction is required for travel to take place;

i) part-circling (with one or more than one directional orientation) – for example, a roll from a top bar to a bottom bar travels but is not a complete circle;

j) combinations of the above – these may be simple or complex.

All of the above categories may be performed with or without turning around specified axes, for example, jumps may be performed with or without turn, sliding down a bench may be performed with or without turning, and so on.

National Curriculum requirements make reference to this aspect of gymnastics in requiring that pupils be taught different means of rolling, swinging, jumping, climbing, travelling on hands and feet (Key stage 2) and, later, to refine and increase range of actions involving travelling through stepping and rolling, flight and sliding (Key stage 3).

Whole body actions – non-travelling

Unlike the actions described in the previous section, the following are performed without travel, on one spot.

a) circling – the body does not travel but moves round a fixed axis, for example, a back hip circle round a bar;

b) pivoting from balance position – often used to lead into travel, for example, fall to prone, or one leg start, pivot forward (followed by forward roll);

c) step turning in place – pirouetting on the hands is an excellent example of this where the body part is removed and replaced without travel; similarly step turns may be performed using the feet;

d) spinning – like sliding, the same point of contact is retained and the body rotates using reduction in friction;

e) swing turning – for example, on a rope or from a single hand reverse hand on a bar;

f) free flight turning – twisting somersaults, turning jumps performed on the spot;

g) lifting – for example, from a headstand lift to handstand;

h) lowering – many situations where the body is lowered, for example, from a head or handstand;

i) tilting/leaning/swaying – a feature of much Modern Rhythmic work;

j) waving – sometimes seen in women's competitive floor exercises; also seen often in dance where a body wave is used to, for example, rise to the feet;

k) swinging – a pendulum swing for example, on a bar with no travel, often as preparation for something else;

l) rocking – also often as a preparation for rolling and to build momentum;

m) springing in place – all jumps on the spot, somersaults on the spot, spotted flic flacs and so on;

n) simple and more complex combinations of the above.

National Curriculum requirements make reference to this aspect of gymnastics on requiring that pupils be taught to refine and increase a range of actions involving turning, flight, spinning, swinging, circling, lifting and lowering the body (Key stage 3).

Combinations of the previous three sections

Actions of body parts

Particular body parts or segments play a crucial role in making certain things possible and in making modifications to actions, for example, the actions of various body parts open up the possibilities of many different endings to a forward roll. If in rolling round a bar both hands are placed on the same side of the body instead of the more usual one on each side, the action will be modified significantly. They may be classified into types of action as follows. The functions of the body parts are discussed in the next section.

a) bend, stretch/straighten type;

b) open, close type;

c) circle, part circle type;

d) twist type;

e) elevation/depression (shoulders);

f) blends of the above.

Additionally, static muscle contractions are especially significant to the maintenance of various whole body and part body positions.

Functions of body parts

These occur in the context of whole body positions and/or actions. The actions of body parts listed in the previous section serve various purposes. They may:

a) initiate, maintain, accelerate, decelerate, redirect, arrest motion (thereby determining the flow and rhythm of motion);

b) change direction, pathway, plane and level of motion of whole body and particular body parts;

c) change relationship of whole body to floor, apparatus, other gymnasts (includes effect on extension of body and on the planes in which motion and stillness occur);

d) change relationships of gymnast's body parts (i.e. determine body form and shape).

They are best considered as:

★ body parts in contact (or about to make contact) with the floor, apparatus, other gymnasts, which maintain, increase or decrease applications of force through actions from actions of body parts a–f to cause: gripping, releasing, supporting, sliding, pushing, pulling, otherwise imparting impetus (for example, to rope). For instance, if, from a headstand, an even push is applied by both hands, the result will be a symmetrical forward roll. If, however, one hand pushes harder than the other the result will be quite different;

★ body parts free from contact (or about to break contact) with floor apparatus, other gymnasts. They will often work by moving the centre of gravity through the action of a body part away from the supporting base. For example, from an arabesque or Y scale, the movement of an arm or the free leg can alter the position of the centre of gravity and thereby initiate a roll forwards. Body parts may:

- make contact to receive and/or support;

- change body form thereby affecting the position of the body's mass centre and the position of the vertically acting line of gravity (sometimes thereby causing gravity to set up or assist the motion of the whole body);

- initiate force in particular body parts and then, by sudden checking, cause their momentum to be transferred to the whole body;

- lead the whole body in particular directions through the actions given in actions of body parts a–f.

Spatial aspects

Spatial aspects are concerned with where action takes place. Appreciation of spatial aspects of movement, while not fundamental in the way that action itself is, can nevertheless enhance gymnastics performance and understanding in various ways:

a) safety – at a very basic level, the ability to share and maximise the use of the available working space is necessary for safe performance;

b) variety – consideration of spatial factors can help to widen the gymnast's vocabulary by drawing attention to new possibilities, for example, performing the same skill in a variety of directions, or at different levels;

c) precision – movements can be performed with much greater accuracy and, in some cases, more successfully if the gymnast is able to judge where in space actions or parts of actions are to take place. For example, the placing of the feet near to or far away from the hips has a fundamental effect on the difficulty of standing up at the end of a forward roll. The ability to judge height and distance is necessary for success in much apparatus work;

d) balance in composition – spatial aspects may complement action considerations in the composition of sequences. For example, a floor exercise will be enriched by a balanced use of different levels, directions and pathways.

These may be achieved through focus on various aspects of spatial orientation.

1 Orientation in personal space, that is, the space immediately surrounding the body – effective use of personal space is related mainly to precision, in that an awareness of exactly where to direct the movement of the whole body or of body parts will lead to much greater accuracy and to more harmonious and aesthetically satisfying movement. The placing of the hands well ahead of the feet in preparation for a handstand or handspring is an example of precise placing in space, improving the efficiency of the skill being performed.

2 Orientation in general space – general space refers to the available working environment, usually gymnasium or hall for gymnastics. Effective use of this space is necessary for safety especially in the class situation where a restricted area may well have to be shared by a relatively large number of pupils. Pupils have to be taught to use the available space sensibly and efficiently, first for safety, and then in such a way that all are able to achieve the maximum amount of work possible.

3 Use of different directions – direction may be changed either by turning to face a new way and then moving off in that direction, or by facing the same way while moving forwards, backwards or sideways. The versatility which the ability to change direction brings can enhance safe use of space and will also enrich the movement vocabulary by adding variety.

4 Use of different levels – high, medium and low levels are all used in gymnastics, although high and low are probably more important than medium. The essence of gymnastics is more likely to be captured in actions which use agility to move from high to low and vice versa. Change of level is an important element in creating interesting and dynamic movement.

5 Appreciation of height and distance – these aspects of space are mainly related to judgements which have to be made in placing and use of apparatus.

6 Use of varied pathways – in space may be related to:
a) floor pattern;
b) air pattern.
Floor pattern refers to the track or pathway made on the floor. It is seen clearly in the patterns followed in artistic gymnastics floorwork. Attention to floor pattern in group work can help to create sequences which enable all members of the group to perform without hindering or blocking the efforts of others.

Pathway of the body in space can be observed by following a body part. Analysis of this kind of pathway is often carried out by recording the position of the centre of gravity throughout a movement. Awareness of the pattern through space may help in skill learning, for example, understanding that a flic flac should follow a long low pathway not a short high one.

Dynamic aspects

Dynamics are concerned with how actions are performed. Factors such as speed and timing of effort are critical to success in many gymnastics skills. The combination of speed and timing of effort are shown very clearly in competitive gymnastics vaulting. Timing can also be seen in work on the high and asymmetric bars. Both are relevant at a much more basic level.

a) Speed – actions may be performed at varying speeds. Performing quickly facilitates some actions and hinders others. For example, the beginner will find a cartwheel easier to perform fairly quickly than very slowly. On the other hand, a headstand is most easily achieved by moving into it very slowly in the early stages.

b) Timing – timing can be a substitute for strength in some actions; in others it is critical for success. For example, headstand lift to handstand can be achieved simply through strength: with timing of effort the same action can be performed with the use of relatively little strength.

c) Use of appropriate force or effort – the beginner frequently feels the need to put in maximum effort in order to complete a movement. As the gymnast gains in experience, the appropriate amount of effort can be utilised, in the appropriate direction. For example, beginners learning a backward roll will often misdirect their effort by shooting their legs upwards instead of keeping a tucked shape which directs the effort over to the floor. Early attempts at a flic flac are frequently misdirected.

Detailed understanding of the dynamic aspects of movement requires a knowledge of biomechanics which is outside the scope of this book. *Gymnastics – A Mechanical Understanding* by Tony Smith goes into various elements of biomechanics in some detail.

Relationships

Relationship may be analysed in terms of floor, apparatus or other people. It may also be considered in relation to action, spatial and dynamic aspects of gymnastics. Its consideration may enrich and add variety to individual work. Partner and group work, where relationship to other people is central, opens up a whole new range of possibilities. It is seen in competitive gymnastics in sports acrobatics. In curriculum work the assistance of partner or groups may enable gymnasts to experience movements beyond their ability when attempted alone.

a) relationship to floor includes:

* ★ right way up or inverted;
* ★ front, back or side facing.

b) relationship to apparatus includes:

* ★ front, back or side facing;
* ★ above, below or to side of apparatus;
* ★ right way up or inverted.

c) relationship to partner or group includes:

* ★ passive (support, obstacle) or active;
* ★ facing, back to back, side by side and so on;
* ★ moving in unison or in canon;
* ★ performing similar or contrasting actions.

Moving in unison

CHAPTER 3

Gymnastics tasks

The selection of tasks which follows is presented in groups based upon chosen elements from the classification in Chapter 2. All are relevant to the National Curriculum programme of study. Because of the way in which the programme of study is written, a number of tasks will be relevant to more than one part of it. For example, rolling activities can be part of turning, part of travelling by rolling or part of flight (e.g. dive forward roll). Cartwheeling can be part of turning or flight (dive cartwheel) in Unit A or of wheeling in Unit B. Some tasks will have been set during earlier key stages – at Key stage 3 the outcomes and expectations will be different, although there will be scope for pupils who have not covered Key stage 2 material to work at a level appropriate for them. Expectations may be more advanced in terms of the level of difficulty anticipated, in the quality of the performance or in the ability of pupils to work independently to answer the task. This chapter should be used in conjunction with the guidance on planning which is given in Chapter 7 so that tasks are selected which will offer challenge, balance and progression for all pupils.

The level of specificity varies from task to task, however no particular teaching style is implied. For example, one of the rolling tasks suggested is 'Roll to different finishes'. This could be presented to the class in several different ways as follows:

1 The teacher could demonstrate a number of different finishes which the class then copies.

2 The teacher could provide workcards with different finishes for pupils who work in pairs helping each other to perform them.

3 The teacher could make a series of workcards available, each showing finishes at different levels of difficulty. Pupils could then choose the finish(es) which they wish to practise.

4 The teacher could ask the class to find as many different finishes as possible.

See Chapter 5 for further information about teaching styles.

The language used here is not intended for presentation to pupils. Appropriate language and terminology will vary according to the age and ability of pupils and according to the preferences of the teacher. Effective teaching is more likely where the teacher has a rich and varied terminology on which to draw.

All the groupings of tasks which follow include material which should have been taught to pupils before they arrive at secondary school. Earlier tasks in each category could therefore be expected to serve as quick revision for the Key stage 3 pupil or as an opportunity for further refinement in the performance of a familiar activity. It has to be recognised, however, that many pupils will arrive at Key stage 3 with very limited gymnastics experience given the number of activity areas which have to be included at Key stage 2, the limited facilities available in some primary schools, often combined with large classes, and the limited access to appropriate INSET for many primary teachers. Tasks should be chosen which ensure that pupils are introduced effectively to new work, that they have opportunities to explore the possibilities, that they are able to select and refine chosen actions and that they are able to use them in the composition of sequences. At Key stage 3 the introduction and exploration phase will already have been completed in many areas and tasks should focus on refining and adding new actions to a personal repertoire followed by sequence and compositional work.

Balance

Balance implies holding a position of stillness. Weightbearing is an elementary aspect of balance involving supporting the body's weight but implying the need for rather less skill than that needed to hold a balance. A balance is often described as referring to the holding of an unstable position. This section includes both weightbearing and balance. In the context of a curriculum relevant to all from the most able to those with special educational needs which might include physical disabilities, it is important that categories are not defined in a way which excludes significant numbers of pupils from achievement. Where pupils are experienced and competent, all the tasks given can be made more challenging by specifying a limited choice of balancing skills as acceptable answers, thus pushing pupils to refine and increase their range of actions.

Balance work includes taking weight on:

★ large body surfaces (front, back, side, hips);

★ combinations of small body parts (two hands plus one foot, two knees plus one hand, etc.);

★ two body parts (hand and foot, knee and foot, two hands, etc.);

★ single body parts (one foot, one hip, etc.);

★ positions where the body is supported below or to the side of the supporting base (hanging, gripping on apparatus, etc.) or between bases (between beams);

★ moving into a balance;

★ moving out of a balance.

Tasks

Choice of small apparatus such as benches and mats or larger apparatus such as beams, vaulting apparatus, etc., should be made as appropriate for the experience and aptitude of the pupils and for the task chosen. For example, balancing inverted would require small apparatus initially but would be within the capability of many pupils using much higher apparatus given confidence and expertise. Balancing beneath a point of support is only possible using certain types of apparatus, such as a beam.

INITIAL TASKS

★ Balance on many different body parts.

★ Balance on large parts/surfaces.

★ Balance so that there are four points of contact with the floor.

★ Balance so that there are three points of contact with the floor.

★ Balance so that there are two points of contact with the floor.

★ Choose one balance and practise it.

★ Balance and then roll until you can repeat the same balance.

★ Choose one balance (for example, shoulder balance, headstand) and find three different body shapes keeping the supporting base the same.

★ Roll to finish balance on four/three/two body parts.

★ Roll, balance and roll out of balance.

★ Jump, land and balance.

★ Jump, land, balance and roll out.

★ Find three balances where part of your body is twisted.

★ From the twisted balance either twist further until you can move out of the balance or untwist until you can move out of the balance.

★ Balance, twist out of the balance, to arrive on your feet and jump showing a twisted shape in the air.

PROGRESSION TO SEQUENCE WORK

★ Link three variations on the same balance using stepping and rolling actions.

★ Link three variations on the same balance using twisting and turning actions.

★ Link together one balance on a large body surface, one on four points of contact and one on two or a single point of contact.

★ Choose three balances and link them using a roll, a jump and a spinning or sliding action.

★ Choose one balance and perform it three times in a sequence using a different action each time as a linking movement.

PROGRESSION TO APPARATUS

★ Find balances so that you are wholly supported on the apparatus (box or bench).

★ Find ways of balancing so that you are supported partly on the floor and partly on the apparatus.

★ Find new balances which are made possible by the apparatus.

★ Balance on bench, travel along bench, balance on bench and roll into a finishing balance on the mat.

★ Find ways of balancing upside down using a rope or ropes so that all or part of your weight is supported on your hands/arms.

★ Find a way of supporting yourself the right way up using a rope or ropes so that your arms/hands are taking all your weight.

★ Balance so that you are upside down.

★ Balance so that you are hanging from apparatus.

★ Balance on apparatus, step out of balance and into a second balance on the mat.

★ Balance on apparatus, roll out of balance and into a second balance on the mat.

★ Balance on apparatus, twist out of balance and into a second balance either on the apparatus or on the mat or floor.

★ Travel across the mat using stepping or rolling actions and balance so that you are supported partly by floor and partly by apparatus.

★ Roll to finish balanced on, under or between pieces of apparatus.

★ Choose one balance on apparatus, one on floor and one using both. Link them to form a sequence.

★ Choose two different balances on each piece of our apparatus arrangements. Practise them.

★ Link one balance on each piece of apparatus to make a sequence.

★ Choose one balance where you are upside down, one using hands and feet and one of your choice and link them to make a sequence.

PROGRESSION TO PARTNER/GROUP WORK

★ Find balances so that you can match your partner.

★ Find balances where you mirror your partner.

★ Link three balances together so that you and your partner synchronise actions and timing.

★ Link three balances together so that you can perform in canon.

★ Link three balances together so that your actions and balances are different but the timing matches.

★ Adapt the above tasks for three or more people.

★ Using apparatus, find two ways of balancing so that the whole group can balance individually and safely at the same time.

★ Link the two balances together so that the whole group can perform at the same time.

★ Add further balances to create a group sequence.

★ Work with a partner to make a balance on hands more stable.

★ Work with a partner to make a balance on one leg (or other body part) more stable.

★ See partner work tasks related to balance.

National Curriculum programme relevance

★ Refine and increase range of gymnastics actions involving balancing skills including the ability to move smoothly into and out of balance (Key stage 3, Unit A);

Balances matching or mirroring partner

★ refine and increase range of gymnastics actions involving twisting (Key stage 3, Unit A);

★ factors that influence quality in gymnastic performances including extension, body tension and clarity of body shape (Key stage 3, Unit A).

Rolling

Rolling includes:

★ skills of forward roll; backward roll; sideways roll; fish/chest roll; circle roll; other activities (see workcards);

★ rolling from different starting positions, e.g. forward roll – from two feet standing, one foot standing, crouch, straddle, arabesque, handstand; backward roll – from crouch, two feet standing, one foot standing, long sitting, straddle;

★ rolling to different finishes, e.g. forwards to two feet, one foot straddle lie flat, V sit, splits, into handstand;

★ rolling from one starting position to a different finishing position, e.g. crouch to one foot stand, straddle to splits;

★ rolling at different speeds;

★ rolling onto, off, along, over, round apparatus;

★ other specific actions: backward roll to straddle, forward roll to straddle, back roll through handstand, dive forward roll.

Tasks

INITIAL TASKS

★ Roll to finish standing on two feet.

★ Roll in a different direction to finish standing on two feet.

★ Roll to finish standing on one foot.

★ Roll to a different finishing position (see workcards).

★ Roll and twist at the end to arrive in a different finishing position.

★ Roll sideways so that only your shoulders make contact with the floor.

★ Find a different starting position and roll to finish standing on two feet (see workcards).

★ Choose a starting position and roll to finish in the same position.

★ Roll starting in one position and finishing in a contrasting position.

★ Jump, land and roll forwards.

★ Half turn jump, land and roll backwards.

★ Roll followed by stretch jump.

★ Roll followed by turning jump.

★ Full turn jump, land roll, full turn jump.

★ Roll, spin and step out of roll.

★ Step, roll and spin.

★ Perform the same roll several times alternating slow, quick, slow.

PROGRESSION TO SEQUENCE WORK

★ Plan and perform a sequence which includes three different rolls.

★ Plan and perform a sequence in which you use a roll of your choice three times.

★ Link three different rolls with three different balances.

★ Link three different rolls with three different jumps.

★ Roll to arrive on a bench, balance on the bench and roll off in different direction. Repeat to form a sequence.

PROGRESSION TO APPARATUS

★ Start on apparatus and roll onto mat.

★ Start on mat and roll to finish on apparatus.

★ Roll along apparatus.

★ Jump off apparatus, land and roll on mat.

★ Come off apparatus in a different way, land and roll on mat.

★ Roll to arrive at apparatus, balance on it, and roll off in a different direction.

★ Travel along apparatus and roll onto mat.

★ Step to get onto apparatus, roll round it, get off and roll on mat.

★ Roll to arrive at apparatus, get on using hands and feet, roll round or between, and get off.

★ Use apparatus to show rolls in three different directions linked to form a sequence.

PROGRESSION TO PARTNER/GROUP WORK

★ Find rolls which you and your partner can perform as matching actions.

★ Find rolls which you and your partner can perform as mirroring actions.

★ Start close to your partner (side by side, back to back) and roll away.

★ Start apart from your partner and roll towards to finish close together.

★ Choose a rolling action and perform it in canon with your partner.

★ In threes or fours, choose a rolling action and perform it in canon using different formations.

National Curriculum programme relevance

★ Refine and increase range of gymnastic actions involving turning, travelling by rolling (Key stage 3, Unit A);

★ refine a series of gymnastic actions into increasingly complex sequences that include variety, contrast and repetition, using both the floor and the apparatus, working alone and with others (Key stage 3, Unit A);

★ factors that influence quality in gymnastic performances including extension, body tension and clarity of body shape (Key stage 3, Unit A).

Taking weight on hands

This aspect of skill is so fundamental to gymnastics work that it is suggested that it should be a part of every lesson. It includes:

★ handstands;

★ cartwheels;

★ bridges;

★ dive rolling activities;

★ walkovers;

★ balancing using hands as support (headstand, half lever, planches);

★ headsprings;

★ handsprings;

★ flic flacs;

★ vaulting activities;

★ climbing activities;

★ heaving and pulling activities;

★ hanging, swinging and circling activities.

For many pupils, the possibilities working on the floor will be exhausted long before apparatus possibilities are exhausted. Many floor skills involving taking weight on the hands are either too advanced for most pupils or require one-to-one teaching and support which is not available in the class context. Pupils who cannot support their weight on their hands in a handstand on the floor, for example, may well be able to with the support of apparatus, such as a box. Partner work will also create opportunities for pupils to help each other and thereby extend the possibilities open to them.

Tasks

INITIAL TASKS

★ Balance so that you use your hands as part of the supporting base.

★ Handstand from single take off or double take off.

★ Handstand, twist to come down in a different finishing position.

★ Roll, stand up and handstand.

★ Handstand into roll.

★ Cartwheel.

★ Begin on feet, take weight on hands and bring feet down in a different place.

★ Turn round or over so that weight is taken on hands and feet only.

PROGRESSION TO SEQUENCE WORK

★ Link three jumps with three actions where your weight is at least partly on your hands.

★ Plan and perform a sequence including one balance and two actions where your weight is on your hands.

PROGRESSION ONTO APPARATUS

★ Balance using hands only, or hands as part of supporting base.

★ Balance with hands on the floor and some other body part on the apparatus.

★ Balance on apparatus and twist so that you can take your weight on your hands to get off.

★ Get onto apparatus taking weight on hands.

★ Get off apparatus by putting your hands on the floor and taking your weight on them.

★ Get off apparatus by taking weight on hands and twisting off.

★ Get over or through apparatus taking weight on hands with and without twist.

★ Get onto apparatus taking weight on hands, travel along and get off taking weight on hands.

★ Travel along bars using hands only.

★ Travel along bars using hands and feet.

★ Travel up ropes using hands and feet.

★ Travel up ropes using hands only.

★ Balance between two ropes supporting weight on hands.

★ Balance underneath a bar taking weight on hands.

PROGRESSION TO PARTNER WORK

★ Find balances with your weight on your hands which you can match with a partner.

★ Find actions taking your weight on your hands which you can match with a partner.

★ Find actions taking your weight on your hands which you can mirror with a partner.

★ Copy your partner's action including weight on hands so that you perform in canon.

★ Balance with or against a partner so that one of you takes your weight on your hands.

★ Help your partner to hold a balance on hands which they could not hold on their own.

★ Balance against your partner so that your weight is at least partly supported on your hands with your partner:

a) standing;
b) lying down;
c) sitting.

★ Get over a partner without contact so that you take your weight on your hands at some point.

★ All the above using apparatus.

★ Adapt the above to work in threes or fours.

National Curriculum programme relevance

★ Refine and increase range of gymnastics actions involving balancing skills including the ability to move smoothly into and out of balance (Key stage 3, Unit A);

★ refine and increase range of gymnastics actions involving twisting and turning (Key stage 3, Unit A);

★ refine and increase range of gymnastics actions involving travelling by stepping (Key stage 3, Unit A);

★ refine and increase range of gymnastics actions involving flight (Key stage 3, Unit A);

★ factors that influence quality in gymnastic performances including extension, body tension and clarity of body shape (Key stage 3, Unit A);

★ refine through practice range of increasingly advanced gymnastic actions involving wheeling (Key stage 3, Unit B);

★ refine through practice range of increasingly advanced gymnastic actions involving swinging, circling, lifting and lowering the body (Key stage 3, Unit B).

Balance against a partner so that one has weight on hands

Flight

Flight includes:

★ jumping with different take offs – single, double;

★ jumping with different landings – to one foot, to other foot, to both feet;

★ jumping for height;

★ jumping for distance;

★ jumping showing different body shape – tuck, pike, straddle, star, stretch, twisted;

★ jumping with turn – quarter, half, full;

★ in different directions – forwards, backwards, sideways;

Balance against a partner so that one has weight on hands

★ onto, off or over apparatus;

★ flight onto hands – dive cartwheel, dive forward roll, flic flac;

★ flight from hands – donkey kick, handspring;

★ flight to catch apparatus – rope(s), high bar;

★ flight to get over apparatus – vaulting activities: through vault, straddle vault, thief vault, overswing vaults – which may involve flight on, flight off or both.

Tasks

INITIAL TASKS

★ Jump with different take offs and landings (one foot to two, two feet to two feet, etc.).

★ Jump showing different body shapes (tuck, pike, straddle, star, etc.).

★ Jump so that you show a twisted shape in the air.

★ Jump so that you turn in the air (quarter, half, full turn).

★ Practise actions where you can show flight from feet onto hands (catspring, dive rolls, dive cartwheels).

★ Practise actions where you can show flight from hands to feet (donkey kick, head spring, handspring, arab spring).

PROGRESSION TO SEQUENCE WORK

★ Plan and perform a sequence using four different jumps to alternate with other actions.

★ Plan and perform a sequence including four different flight actions.

★ Plan and perform a sequence including use of flight to negotiate partner and flight within linking actions.

PROGRESSION TO APPARATUS

★ Jump to arrive on/under/at apparatus (from springboard onto box/horse; to catch bar or beam (use diagonal approach and split grasp if using beam); to catch still rope; to catch moving rope; to catch two still ropes; to catch two moving ropes.

★ Jump off apparatus (with control; with different body shapes; with turn).

★ Get onto apparatus and get off taking weight on hands with flight (gate vault; squat on, straddle off box; headspring off box; catch and release ropes, etc.).

★ Get over apparatus so that only hands contact it (longfly over long box; through vault; straddle vault; long arm overswing; short arm overswing; flank vault; thief vault; other ways of getting over vaulting apparatus using hands, e.g. with quarter turn; vault over single beam.

★ Vaulting activities:
Most vaulting activities involve taking the weight on the hands and can be taught to some or all of a class. Appropriate support from the teacher must be given where necessary. Pupils can be taught to support each other. Decisions as to how much pupils are to be given the responsibility for this must be made by the individual teacher. The pupil's capacity for taking responsibility in this way is often underestimated. It does require the development of the pupil's ability to take responsibility and the cultivation of an appropriate working atmosphere.

PROGRESSION TO PARTNER/GROUP WORK

★ Find four different jumps which you can synchronise with your partner.

★ Find other actions involving flight which you can synchronise with your partner.

★ One balance and the other use a flight action to get over partner.

★ Choose a flight action (jump, vault) and appropriate apparatus and perform in canon.

★ Develop the above task to increase group size and vary approach angles.

National Curriculum programme relevance

★ Refine and increase range of gymnastics actions involving flight (Key stage 3, Unit A);

★ factors that influence quality in gymnastic performances including extension, body tension and clarity of body shape (Key stage 3, Unit A);

★ refine through practice range of increasingly advanced gymnastic actions involving wheeling (Key stage 3, Unit B)

Partner/group work

Partner work may be used as a development of work on other topics as indicated in previous sections of this chapter or as an area of work in its own right. Partner work may be with or without contact with the partner. The latter clearly demands greater co-operation and interdependence and will generally involve more advanced work because of this.

Partner work may include:

★ matching and mirroring partner's actions;

★ different relationships to partner (side by side/facing/back to back, etc.);

★ meeting and parting;

★ assisted balance – one active, one passive;

★ counter balance and counter tension;

★ partner balances – one base, one balancer;

★ assisted flight.

For some partner work, partners need to be of fairly similar height and weight.

Non-contact partner work

TASKS
★ Perform any action so that you copy your partner exactly, starting side by side and synchronising actions.
★ Do the same, but start opposite partner. Do the same but start back to back. Do the same starting side by side but facing opposite directions.
★ Find three more actions that you can both perform and perform these actions as in the previous three tasks.
★ Create a sequence with your partner including at least four matching actions.
★ Create a sequence with your partner including at least four matching actions showing changes of speed and different relationships to your partner.
★ Choose three different actions and perform them in canon, that is one after the other.
★ Choose four actions and perform them so that you alternate synchronisation with canon.
★ Create a sequence with your partner which includes both canon and synchronised actions.
★ One balance so that partner can jump over.
★ One balance so that partner can get over or under in other ways (without contact).
★ One balance showing a large or small body shape. Partner show a similar body shape when negotiating obstacle.

- ★ Do the same but show a contrasting body shape.
- ★ After negotiating partner, find a balance so that partner can get over you.
- ★ Travel to partner, balance together and travel away.
- ★ Create a sequence with your partner so that you alternate balancing with negotiating your partner's balance. Use a variety of actions as linking movements.
- ★ Using box/horse, beams or ropes, adapt balances so that you create a barrier for your partner to negotiate on/over or through the apparatus.

Contact partner work

TASKS

- ★ One help partner to hold a balance which he or she could not hold alone.
- ★ Experiment with counter balance positions with a partner (see workcards).
- ★ Choose a counter balance and find a way of moving into and out of it using different actions.
- ★ Experiment with counter tension positions with a partner (see workcards).
- ★ Choose a counter tension position and find a way of moving into and out of it using different actions.
- ★ Create a sequence with your partner using counter balance and counter tension positions with a variety of linking actions.
- ★ Adapt some of your counter balance or counter tension positions to apparatus.
- ★ Adapt some of your counter balance or counter tension positions to working in a group of three or four.
- ★ Balance so that you are taking your partner's weight (see workcards).
- ★ Move towards your partner, balance and move away.
- ★ Repeat with other balances.
- ★ Create a sequence so that you include four partner balances and a variety of linking actions.
- ★ Choose a balance and adapt it for three or four people.

★ Create a sequence including balances for two and three people and a variety of linking actions.

National Curriculum programme relevance

★ Refine a series of gymnastic actions into increasingly complex sequences that include variety, contrast and repetition, using both the floor and the apparatus, working alone and with others (Key stage 3, Unit A);

★ develop, refine and evaluate a series of actions, with or without contact with others (Key stage 3, Unit B).

Sequence work

Sequence work has been included as a progression for all of the tasks already provided. Further development of sequence work involves:

★ increased length and complexity;

★ refinement of performance quality;

★ further requirements related to variety, such as different speeds, levels, actions, formations;

★ development of criteria for assessing performers' sequences and those of their peers;

★ application of the criteria developed.

TASKS

★ Plan and perform a sequence which follows the pathway drawn on the chalkboard and which includes the actions specified.

★ Plan and perform a sequence to include the following:
a clearly defined pathway which includes both straight lines, curves and angles;
continuity and flow;
a variety of actions including some where the body is upside down;
clear changes of speed.
Evaluate your sequence and that of another group.

★ Plan and perform a sequence to include the following:
at least four partner balances;

smooth transitions into and out of balances;
a variety of actions between balances;
links which use both canon and unison.
Evaluate your sequence and that of another group.

★ Plan and perform a sequence which uses all the apparatus to include the following:
one group balance involving everyone;
the involvement of everyone in one partner balance;
each member of the group using all pieces of apparatus;
variety and contrast.
Evaluate your sequence and that of another group.

National Curriculum programme relevance

★ Refine a series of gymnastic actions into increasingly complex sequences that include variety, contrast and repetition, using both the floor and the apparatus, working alone and with others (Key stage 3, Unit A);

★ develop, refine and evaluate a series of actions, with or without contact with others (Key stage 3, Unit B).

4

Gymnastics skills

The teaching of gymnastics skills is frequently equated with a direct class activity approach and criticised as an educational experience because of the high skill threshold of many gymnastics actions. There are, however, a number of core skills which arise in gymnastics work, whatever approach has been adopted, and which may be practised with benefit by a whole class. There are others which the teacher can incorporate into curriculum work provided that he or she has the facility to utilise teaching approaches which enable pupils to work in groups on differentiated tasks. This offers the possibility of including work on more advanced skills for one or two groups within a class. At Key stage 4, where numbers may be quite small, especially within, for example, a GCSE option group, more of this work may be included, although it has to be remembered that, by Key stage 4, many pupils are past the optimum age for learning such skills.

This chapter looks at some common core skills, at the teaching points which should be considered, at common faults which may be found and at progressions and variations which will be needed in order to continue to provide a challenge for the mixed ability class.

Forward roll

Teaching points

★ crouch start;

★ hands flat on floor about shoulder width apart, fingers forward;

★ push with legs, tuck head well in;

★ weight onto back of head and shoulders;

★ keep tucked, round back;

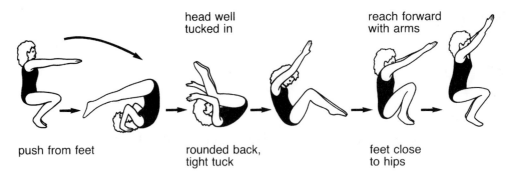

Fig 4.1 Forward roll

★ put feet on floor close to hips, reach forward and up with arms;

★ when weight is on balls of feet, stand up.

Common faults

★ head not tucked in – discomfort in roll;

★ back not rounded – discomfort in roll;

★ hands incorrectly placed – failure to stand up or failure with later skills;

★ feet too far away from hips – failure to stand up;

★ uneven push from hands – crooked roll;

★ head turned sideways – crooked roll;

★ insufficient push from legs – failure to roll.

Progressions

★ rolling to different finishing positions (see workcards);

★ rolling from different starting positions;

★ dive forward roll;

★ rolling off apparatus;

★ rolling on apparatus, e.g. along benches, flat surfaces;

★ rolling onto apparatus, e.g. onto boxes or other flat surfaces;

★ rolling round beams, bars.

Backward roll

Teaching points

★ crouch start;

★ hands in position early, flat on floor close to head, thumbs point towards ears;

★ tuck tightly, knees on chest;

★ push hard from hands, as hips pass over head, stay tucked;

★ maintain push and land on balls of feet.

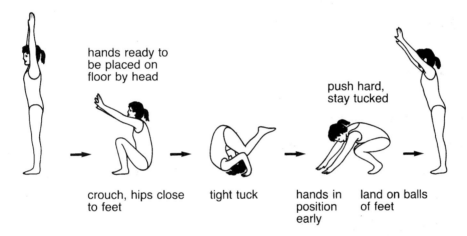

Fig 4.2 Backward roll

Common faults

★ lack of momentum – unable to complete roll;

★ hands incorrectly placed – unable to push;

★ uneven push from hands – crooked roll;

★ opening out too soon – unable to get to feet or to get over at all.

Progressions

★ rolling to different finishing positions;

★ rolling from different starting positions;

★ straight leg roll through long sit position;

★ rolling on apparatus;

★ rolling off apparatus, for example, box, beam;

★ back roll through handstand.

Practices for rolling

★ rocking to and fro, staying tucked and building momentum;

★ rolling on a slightly inclined surface;

★ rolling between ropes;

★ rolling from box, forwards.

Headstand

Teaching points

★ head and hands in triangle on floor;

★ forehead on floor;

★ walk feet up towards hands, on toes;

★ push down with hands and lift feet into tucked headstand;

★ keep back straight;

★ straighten legs.

Common faults

★ head and hands in line on floor – overbalance;

★ back of head on floor – overbalance;

★ straightening of legs too soon – failure to invert;

★ straightening legs when off balance – collapse of headstand;

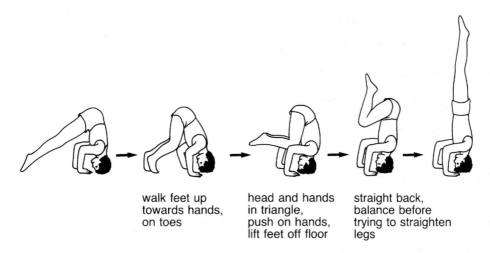

walk feet up   head and hands   straight back,
towards hands,   in triangle,   balance before
on toes   push on hands,   trying to straighten
                      lift feet off floor legs

Fig 4.3 Headstand

★ back not straight – collapse of headstand.

Progressions

★ from different starts;

★ to different finishes;

★ with straight legs throughout;

★ from lying flat on front;

★ off apparatus;

★ on flat surfaces.

Handstand

Teaching points

★ hands shoulder width apart;

★ fingers face forward;

★ eyes look at hands;

★ arms straight;

Fig 4.4 Handstand

★ hips over shoulders;

★ body straight.

Common faults

★ hands too far apart – handstand collapses;

★ arms bent – handstand collapses;

★ head looking too far forward – hyperextension in spine or inability to invert;

★ backarching instead of straight – weak position.

Progressions

★ 'kicking horses';

★ hands on floor, walk feet up wall or wallbars;

★ with support;

★ against wall;

★ from box;

★ from different starts;

★ to different finishes;

★ from two footed take off;

★ from two footed take off with legs straight;

★ off apparatus;

★ over apparatus.

Cartwheel

Teaching points

★ face forward to start;

★ chest towards knee of bent leg;

★ push hard from bent leg, swing other leg up;

★ wide straddle;

★ arms straight;

★ head look at floor;

★ body straight;

★ put first foot close to hands, eyes looking at hand on floor;

★ take first hand off floor just before first foot lands;

★ hands and feet should contact floor in a straight line.

Fig 4.5 Cartwheel

Common faults

★ incorrect sequence of hands and feet;

★ poor body position;

★ legs not lifted over shoulders.

Progressions

★ from different starts;

★ with flight;

★ using one hand only;

★ along flat surface, e.g. bench or box;

★ over apparatus;

★ several in succession;

★ cartwheel with either hand leading;

★ arab spring or round off (cartwheel, bring legs together when vertical and twist to bring both legs down at same time; second hand should be placed so that fingers face first hand).

Headspring

Teaching points

★ hand and forehead placed on floor simultaneously;

★ hips lead body off balance;

★ strong swing from straight legs;

★ extension of hips to arch back;

★ strong push from hands;

★ push hips forward to land on feet.

Common faults

★ poor timing – usually early push – insufficient rotation;

★ lack of hip extension, bent legs;

★ insufficient push from hands – failure to get to feet;

Fig 4.6 Headspring

★ excessive arching of back – over-rotation;

★ lack of body tension – collapsing.

This skill should be taught with support. Many pupils who find it difficult to perform this on the floor can learn this skill as a vault.

Progressions

★ headstand with legs straight through piked position.

Handspring

Teaching points

★ hurdle step preparation – long and low;

★ shoulders extended, chest towards floor;

★ hands shoulder width apart, fingers pointing forward, head slightly back;

★ strong leg kick, arms straight, shoulders behind hands;

★ strong push from hands/shoulders, legs come together;

★ keep body straight and maintain body tension;

★ land on balls of feet.

Fig 4.7 Handspring

Common faults

★ hands placed on floor close to feet instead of reaching forward;

★ bending arms;

★ lack of body tension causing poor body position with insufficient extension;

★ push from shoulders at wrong time;

★ lack of flight with flexible pupils who bend their backs instead of pushing off;

★ tucking or collapsing into landing leading to pitching forward;

★ falling backwards on landing from insufficient momentum caused by lack of swing from legs and/or push from hands.

The pupil must be able to perform a good handstand before attempting this skill. It should be learned with support – even when the pupil is safe without support, it is often needed if technique is to improve.

Progressions

★ off the end of a low box with support;

★ off a reversed springboard with support;

★ for the proficient performer, to one leg or from two feet (flyspring).

Flic flac

Teaching points

★ push on take off through heels, not balls of feet;

★ knees should be behind feet with lower leg vertical;

★ take off from sitting position;

★ strong push from feet backwards and upwards;

★ arms swing forward and up over head;

★ head in line with body;

★ body extended but not excessively so;

★ legs together and straight;

★ body should pass through handstand;

★ bring feet down together while pushing away from floor with hands.

This move must be taught with support.

Fig 4.8 Flic flac

Common faults

★ incorrect angle of take off, drive from toes or balls of feet instead of heels;

★ drive taking body high rather than along – flic flac should be long and low;

★ arms not in position for landing;

★ tucking legs up or failing to keep legs together;

★ twisting sideways on take off.

Progressions

★ handstand to bridge and stand;

★ from standing backwards into crab position;

★ with support from slight downward slope – reversed springboard, inclined benches – sufficient matting must be provided;

★ once the performer is proficient, following an arab spring;

★ two or more consecutive flic flacs;

★ with split legs (gaynor flick).

Through vault

Teaching points

★ two foot contact with springboard with feet ahead of body, arms back;

★ swing arms forward as feet leave board;

★ reach for box top;

★ hands firmly placed on box top, shoulder width apart or slightly wider;

★ straight arms, fingers forward;

★ knees bent to squat position;

★ head up;

★ push up and forward from box as knees are drawn up to chest;

★ swing arms up to achieve a stretch before landing;

★ give with ankles, knees and hips on landing.

Fig 4.9 Through vault

Common faults

★ head down causing over-rotation;

★ mis-timed push from box;

★ insufficiently tight tuck combined with lack of push from arms resulting in feet contacting top of box;

★ feet taken to side often from lack of confidence;

★ leaving hands on box too long or even until landing.

Most of these faults require support by the teacher so that pupils can have the confidence to put them right.

Progressions

★ squat onto box to get knees on top;

★ squat onto box to get feet on top;

★ squat onto box and jump off;

★ once basic vault is achieved, move springboard back to increase flight on;

★ when springboard is far enough away, straight body position during flight onto box;

★ with sufficiently strong thrust from arms, straight body can be maintained throughout.

Straddle vault (widthways box)

Teaching points

★ two foot take off from springboard;

★ arms swing forward at take off and reach for box;

★ hands shoulder width apart, arms straight;

★ head up throughout;

★ legs in wide straddle straight after take off;

★ push upwards and forwards from hands;

★ legs come together to land;

★ give with ankles, knees and hips on landing.

Fig 4.10 Straddle vault

Common faults

★ over-rotation leading to overbalance;

★ too hard a push from feet leading to overbalance;

★ insufficient push from hands leading to overbalance;

★ leaving hands on the box for too long leading to overbalance;

★ straddle position not wide enough for legs to clear box (stronger push from hands can solve this).

This vault needs teacher support to give confidence and for safety.

Progressions

★ on floor, front support position, jump feet to straddle;

★ straddle onto box with teacher to support;

★ vault is easier to perform over a buck;

★ layout position in flight onto box once vault has been achieved with confidence – springboard will need to be moved back;

★ same vault performed over box lengthways, once pupil can vault over a buck with the springboard far enough back to give the length of flight needed to clear a long box. Support should be provided.

Overswings over cross box

Headspring

Teaching points are as for headspring on floor.

Progressions

★ from end of low box;

★ over low box;

★ with increasing height from end of box;

★ with increasing height over cross box using springboard.

This vault must be taught with support. It is not suitable for performance over a long box unless performers are very skilled because of the pressure on the neck if technique is not very good.

Long arm overswing

Teaching points:

★ fast approach;

★ strong thrust from springboard, reach for box with hands;

★ body straight, plenty of muscle tension;

★ strong push from box with hands for lift and rotation;

★ give with ankles, knees and hips on landing.

Fig 4.11 Long arm overswing vault

Common faults

★ box at wrong height: too low and push from legs and hands becomes difficult; too high and pupils cannot succeed;

★ insufficient thrust from legs – failure to get upside down;

★ mistimed push from arms;

★ failure to push off from arms at all;

★ bending back – lack of flight off and over-rotation.

Progressions

★ handstand from end of long box;

★ handspring from end of box.

For all vaulting, if springboard or trampette are to be used it is essential both for safety and for the development of successful technique that pupils are taught to use them correctly.

CHAPTER 5

Teaching styles

A number of different writers have, at various times, identified different teaching styles and related them to particular philosophies of teaching or to specific learning outcomes. All define these teaching approaches in terms of the extent to which pupils are involved in the learning process. Earlier this century, physical education syllabuses gave no freedom of choice to the pupil, and little to the teacher, who was expected to follow the tables of exercises prescribed. These gave precise guidelines not only on the movements required, but also on the instructions which were to be given by the teacher.

The change in philosophy towards a more child-centred approach to education led to changes in method of presentation. Bilbrough and Jones (1973) described approaches to gymnastics teaching as 'direct', 'limitation' or 'indirect'. Kane, in a large-scale study of physical education teaching (Schools Council 1974), used the categories 'direct', 'guided discovery', 'individualised learning', 'problem solving' and 'creative', although no definition of the various styles was offered.

The most detailed analysis of teaching styles is that undertaken by Mosston and Ashworth (1986) who describe a spectrum of teaching styles in terms of the extent to which decisions are taken by pupil or teacher. Although this work is open to criticism, not least for ignoring the context of learning and factors such as teacher expectations, it has become the most widely used framework in both initial and in-service training in this country over recent years. The following description of the various styles is based upon a broad interpretation of Mosston's definitions and indicates ways in which pupil involvement in learning can be increased even within styles which give most of the decision-making to the teacher.

Style A. Command

This style can be seen in a variety of activity contexts. Maybe the most obvious example is in the typical aerobics class which displays all the attributes of the Command Style. Pupils follow instructions, respond as a class, conform in their response, achieve high levels of activity and numerous repetitions of the various exercises. Some gymnastics warm-up routines mirror this. It can also be seen in the gymnastics context in the teaching of high-risk activities where the teacher takes all decisions in the interests of safety. For example, in vaulting activities, where pupils are learning with teacher support, the teacher will not only tell the pupil exactly what to do, but will also dictate the timing. The same procedures will be used in teaching some tumbling activities.

Characteristics

★ teacher takes decisions;

★ pupils are expected to follow instructions and perform;

★ immediate response;

★ large amount of time on task and for repetition;

★ no allowance for individual differences.

When?

★ when conformity is required;

★ when a high level of activity is a priority;

★ when safety is paramount.

Likely learning outcomes

★ conformity;

★ uniform behaviour;

★ potential for high levels of activity;

★ safe learning of high-risk activities and, therefore, confidence;

★ accurate replication of a movement.

Examples of tasks

1 Pupils carry out a class warm-up routine following instructions given by the teacher.

2 Pupils are told to attempt a front somersault on the trampoline on a count of three so that the teacher can support them.

Style B. Practice

While the teacher still takes most decisions about what is to be practised and in what way, pupils begin to take some responsibility. They will be able to choose the timing or the pace of the activity. They may also be given choice of location – 'Find a space where you can practise safely' – or of difficulty – 'Choose one of the balances from the worksheet and practise it'.

Characteristics

★ teacher takes most decisions;

★ pupils take some decisions about, e.g. location, timing, pace;

★ teacher has time to give individual feedback;

★ pupils are able to work individually, at their own pace.

When?

This is probably the most commonly used style when teaching or working to improve basic skills. While its key feature is that the teacher is able to go round giving feedback, the use of other strategies in conjunction with this style will affect the relative emphasis given to rote performance, to knowledge or to understanding. For example, the teacher can choose simply to tell the pupils how to improve – 'Place your feet close to your hips so that you can stand up at the end of the roll' – or can ask questions to make the pupil think – 'Where do you think you need to put your feet to make it easier to stand up? Why?'.

It is also appropriate when the teacher wants to retain a strong control over the activity, perhaps for safety reasons. This could be when teaching the safe use of the trampette or when introducing a new skill in the gymnasium. It might be needed if the teacher felt that pupils were not ready to take responsibility and were not capable of working independently.

It may also be used where the efficient use of limited time to include maximum activity is a priority.

Likely learning outcomes

★ sustained practice leading to improved performance;

★ refined skills;

★ development of new skills;

★ development of concentration and perseverance;

★ development of the ability to practise independently;

★ limited decision-making capability through choice of pace of practice or location;

★ potential for practising at individual level thereby catering to some extent for individual differences.

Examples of tasks

1 Pupils have one bench between three. They are shown how to approach the bench and jump off it as practice for later use of a springboard. They are then asked to practise concentrating on a double take off from the bench and jumping for height.

2 Pupils have a workcard giving examples of different endings for a handstand (or action taking weight briefly on hands). They choose one example and practise it. The teacher goes round giving individual help. She stops the class occasionally to remind them all of the importance of good quality work.

Style C. Reciprocal

The key feature of this style is that pupils work together with one playing the teacher and one the pupil role. Its advantage is that the performer can receive far more feedback than the class teacher can possibly give. In order to help pupils to know what sort of feedback is appropriate, task cards are frequently used with this style. When introducing this style to pupils, it is important to spend time ensuring that all are aware of the qualities needed to play the 'teacher' role effectively, e.g. give praise, be patient, watch carefully, etc. While pupil activity may be reduced pupil learning will not be, since adopting the role of teacher may have a beneficial effect on the pupil's performance when it is his/her turn to be active.

Characteristics

★ pupils work in pairs;

★ one performs, partner gives feedback;

★ pupils receive immediate feedback from partner;

★ teacher communicates through pupil to performer;

★ emphasis on personal and social skills because of demands on communication skills.

When?

★ when personal and social development through pupil interaction is an aim;

★ to foster evaluation skills;

★ when increased feedback for pupils is desirable;

★ to give recognition to different abilities.

Likely learning outcomes

★ co-operative behaviour and development of social skills;

★ development of observational and analytical skills;

★ development of the ability to evaluate others' performance against given criteria or against self-chosen criteria;

★ development of communication skills through giving positive and constructive feedback;

★ practising and refining skills;

★ understanding and evaluating how well others have achieved;

★ ability to help others through suggesting approaches to improvement or giving specific technical advice.

Examples of tasks

1 Pupils, in twos, are asked to practise balancing on either their heads or their hands. Teaching points for both are written on the chalkboard. They take it in turns to practise. Their partner is asked to help them to improve.

2 Pupils, in groups of three, have a workcard giving several examples of counter balances to be performed in pairs. A number of teaching points are given on the workcard. One pupil helps the other two to perform the balances. They change roles so that all have a turn at being both 'teacher' and performer.

Style D. Self-check

For this style to be successful, a certain amount of proficiency and understanding is needed so that pupils are able to make a sensible assessment of their success and appreciate how they might improve further. It is more appropriate to some activities than to others. For example, it is far more difficult for the performer to assess his or her success or to analyse his or her performance if the activity is fast moving and has little reference to external success criteria. Tumbling skills in the gymnasium are far more difficult to self-assess than shots at goal in hockey or basketball, putting the shot or throwing the javelin. For the latter examples the performer can be given clear indicators and guidelines for diagnosing faults. Success in shooting at goal is easily measured, as is the distance achieved in a put or throw. In a gymnasium, the use of mirrors can facilitate the use of this style. Video is another means of enabling pupils to assess their own performance.

Characteristics

★ pupils assess own learning against given criteria;

★ shift towards independent learning;

★ certain level of proficiency needed. Tasks need to be conducive to assessment through reference to 'external' indicators.

When?

★ when pupils are capable of sustained and constructive practice and have developed some skill in evaluating their own performance.

Likely learning outcomes

★ ability to evaluate own performance;

★ development of sustained independent practice;

★ adapting and refining skills;

★ practice and performance.

Examples of tasks

1 Pupils are using springboards and trampettes and are jumping for height. A chalk line is drawn on the landing mat and pupils are asked to try to land no further out than the line.

2 Pupils are encouraged to use mirrors to check body tension and position in specified balances.

Style E. Inclusion

This is the first style which plans specifically for individual differences, although they are not excluded from the other styles mentioned so far, with the exception of the Command Style. The slanted rope is often given as an example of how this style works. A horizontal rope means that all have to clear a uniform height. If the rope is slanted, then each individual can choose the height at which they clear it. Similarly, in other activities, a range of options is given to pupils so that they can participate at a level appropriate for them. For example, pupils can be given beams or vaulting apparatus at varying heights and choose that best suited to their size or ability. Guidance may be needed to help pupils to judge the level at which they should be working. Some may need encouragement to challenge themselves further while others may choose a level for which they are not yet ready.

Characteristics

★ allows for individual practice at appropriate difficulty level;

★ assumes a certain level of self-motivation and awareness of limitations if all are to work to their capacity;

★ tasks are set with varying degrees of difficulty, e.g. a range of gymnastics balances, a choice of activities involving taking weight on hands;

★ pupil makes decision about which option is appropriate for him/her and when or whether to progress.

When?

★ when differentiated opportunities are needed if all are to continue to progress;

★ where pupil choice is important;

★ where independent learning is desired.

Likely learning outcomes

★ differentiation with potential for learning across wide ability range, provided that pupils make appropriate choices;

★ ability to evaluate own performance against given criteria and to compare it with that of others;

★ progression at own pace;

★ responsibility for own learning in choosing level of performance and pace of progression;

★ adapting and refining skills and development of new skills;

★ practice and performance.

Examples of tasks

1 Pupils are given a choice of activities, all involving taking their weight on their hands (handstand, handstand from two foot take off, cartwheel, one handed cartwheel). They choose one to practise and move on to a more difficult challenge once they have mastered their original choice.

2 Pupils are given a choice of apparatus (mat, padded bench, low box, high cross box, high long box). They are asked to practise one of a limited choice of balances (shoulder balance, headstand) on one of the pieces of apparatus, including consideration of how they are going to get into the balance.

Style F. Guided discovery

In this style the teacher still determines the goal but guides pupils to learn for themselves through carefully presented tasks or questions, or a combination of the two. For example, pupils might be asked a series of questions to elicit the appropriate response. 'Balance so that you are very stable and can't be pushed over. What sorts of body parts are you using? What level are you at? Now balance so that you can hold still but so that you could easily be pushed over. What sort of body parts are you using? What level are you at? What do you have to do to keep that position? What happens if you move free body parts?' Through asking relevant questions the pupils can be led to appreciate that body tension is needed to hold difficult balances and that the centre of gravity needs to be over the supporting base. They will have discovered that very stable balances involve large body parts or several body parts and are likely to be fairly low level.

Characteristics

★ teacher systematically leads pupil to discover a predetermined learning target;

★ pupil is involved in a process of discovery based upon questioning by teacher;

★ pupils involved in thinking about answers and learning with understanding;

★ activities need to be appropriate, i.e. this style is not suitable for activities where experimentation would be unsafe. It will take longer than other styles and so should be used where the ability to work out solutions independently will be important in the future.

When?

★ to promote understanding and thought

Likely learning outcomes

★ Understanding of work undertaken – tactics, principles of movement, etc.

Examples of tasks

1 Pupils are asked to try each of several alternatives and decide which is the most effective. 'At the end of a forward roll, try standing up first with your legs straight, and then with them bent. Which is the easier? Try rolling fast and slowly. Which is the easier to stand up from?'

2 Pupils are asked to balance on their heads and work out where head and hands need to be positioned for the most stable base.

Style G. Divergent or problem-solving

This style encourages creative thinking in that no one solution is assumed and the outcome cannot necessarily be predicted, although the experienced teacher will have some idea of the kinds of solutions likely to emerge.

Characteristics

★ teacher presents problem;

★ pupils are encouraged to find many alternative solutions;

★ there are no finite predetermined answers. An infinite number of solutions could be possible;

★ appropriate for activities where creativity is important, such as gymnastics and dance.

When?

★ where the development of new ideas is a priority;

★ to promote understanding and thought;

★ where the ability to think through solutions is important;

★ where time is needed for exploration and trying ideas out.

Likely learning outcomes

★ development of planning and evaluation skills;

★ development of ability to devise solutions;

★ compositional skills.

Examples of tasks

1 Pupils are asked to find different ways of starting or ending a forward or backward roll.

2 Pupils have been taught a number of partner balances. They are asked to work in threes or fours and find balances which can be performed in the larger group.

Style H. Individual programme

Styles H onwards are more likely to be found in higher education or at Key stage 4 and beyond. The individual programme may be found, for example, in a GCSE practical programme where the teacher decides that gymnastics or dance will be an assessed area and pupils make decisions about the focus of a dance or gymnastics routine. Pupils may choose the focus of the project which constitutes part of the course work for GCSE, GNVQ or A level.

Characteristics

★ teacher decides the general area for study;

★ learner takes decisions about the detail of what is to be studied;

★ teacher acts as adviser and facilitator.

Likely learning outcomes

★ project work such as GCSE;

★ Key stage 4 work where pupils have options within their chosen activity areas.

Style I. Learner initiated

This style is more usually identified with independent work in higher education where the student has freedom to suggest topics for a dissertation or thesis and the teacher's role is one of facilitator, responding to questions from the student.

Characteristics

★ learner takes initiative about content and the learning process;

★ teacher/supervisor acts in an advisory capacity when approached by the learner.

Likely learning outcomes

★ independent study, such as work for a dissertation or thesis.

Style J. Self teach

Not applicable in the school context.

Teaching strategies

One of the features of gymnastics in the curriculum should be that it offers opportunities for using varied teaching methods, from open-ended tasks in order to foster individuality and promote thoughtful application to solve problems, to closed tasks set by the teacher in order to improve specific skills or improve quality of performance or increase physical activity.

Whichever general approach is adopted, the teacher will need to select appropriate teaching techniques to reinforce what is wanted and to eliminate the irrelevant. Several different approaches will often be needed in order to make progress to the desired result. These techniques include effective task setting and explanations, good observation, use of demonstrations and questioning, and resource-based teaching.

Task setting

What kind of task is being set? Is it about management, that is setting out equipment, organising pupils into groups, organising them into the working space? It is about performance, that is what to do to learn or improve a particular skill or compose a fluent sequence?

★ Are all pupils listening?

A very common mistake made by beginning teachers is to talk over pupils who are still conducting their own conversations. This is usually the result either of lack of confidence or over-eagerness to get on with the task in hand.

★ Is the language used appropriate? Do all pupils understand?

By the time they reach Key stage 3 some pupils have given up hope of understanding what has been said and rely on watching other pupils and copying. Finding out exactly what pupils do understand can sometimes provide insights into their behaviour. What has been interpreted as misbehaviour and constant off-task behaviour can actually be a result of never understanding the task in the first place.

★ *Do pupils know why they are being asked to complete a particular task?*

Detailed explanations of 'why' are not needed every time a task is set, but pupils should know what they are expected to know, be able to do and understand by the end of any given unit of work. They then have a context within which the various tasks within a lesson are set.

★ *Is the complexity appropriate for the group? Is it too complex? Insufficiently challenging?*

A common mistake is to present pupils with far too much information within a single task and then leave them for far too long working on one complex task. With younger pupils or those whose concentration span is limited, several simply explained tasks interspersed with concentrated activity periods are likely to be more effective than one complicated task leading to pupil attention wandering and activity tailing off.

★ *Does it allow for different abilities?*

Most physical education teaching is to mixed ability groups. It is useful to ask in relation to each task, 'What will I expect the most able to do, the least able to do, the average child to do?'

★ *Is it presented concisely?*

Most analyses of physical education teaching reveal relatively small proportions of active learning time and quite low levels of physical activity over the duration of a lesson. One reason for this is the length of time taken to set tasks and organise pupils. Sometimes the problem is that too much information is being presented. Sometimes the teacher is simply too verbose!

★ *Does it sound interesting and motivating?*

This is related to confidence and to the effective use of the voice. Lack of confidence can lead to a very dull presentation.

Explaining

Many of the factors identified as important for effective task setting are equally applicable to explaining, especially those related to language use, conciseness and complexity. Other questions might be:

★ Could a demonstration replace or supplement the explanation?

★ Could pupils be involved in the talking rather than just the teacher?

Observing

The ability to observe and to react to what is seen is absolutely fundamental to good physical education teaching. It becomes even more important where a large class is working on differentiated tasks, where, if the teacher's powers of observation are limited, they will not be able to develop the work being produced. The following procedure is suggested to help students to adopt a safe and effective routine when pupils are working and one which will facilitate the development of observation skills.

★ Is the class working safely?

This will always be a factor although clearly it is a more significant issue in some activity areas than others. While most pupils tend to avoid attempting tasks beyond their abilities, they may unwittingly be encouraged to do so by peer group pressure or by suggestion by the teacher. Lack of awareness of others may lead to pupils working too close to others for safety or to failing to wait until there is room for them to take their turn safely. Equipment may be positioned so that it becomes unsafe for use, e.g. gymnastics apparatus too close to a wall or window.

★ Is the class answering the task set?

The teacher needs to check this before moving on to anything else. If the task is not being answered, is this because:

a) pupils have not listened?;

b) there is genuine misunderstanding?;

c) the task is too difficult or too easy or otherwise inappropriate?;

d) pupils are deliberately working off-task?

★ How well is the task being answered?

a) Is a variety of response sought? If so, are the ideas being produced relevant to the task set?

b) Is quality of performance being sought through teaching a specific skill or giving the opportunity to practise a selection of actions?

c) Are all pupils performing with a high rate of success? Is this because the task is too easy?

d) Are most struggling? Can you help or does an easier task need to be set?

★ Which aspects of the work need improvement?

a) Are there common difficulties which demand some input to the whole class?

b) Do one or two individuals need specific help?

c) How are you going to phrase the additional input? Remember pupils need to know not only what they need to do, but also how to do it (e.g. rather than say 'Stand up at the end of the roll' say 'Position your feet closer to your hips so that you can stand up at the end of the roll').

★   *Are there some examples which could be shown to the rest of the class?*

Remember to prewarn pupils that you would like them to demonstrate.

★   *Are individuals working to their capacity?*

Has the task allowed for this? If not, how can the able be extended?

Questioning

The skilful use of questions is an essential part of effective teaching. Questions are used for various purposes and it is important that the teacher is clear about why a particular question is being asked so that it can be phrased accordingly. For example, if the teacher wished to check whether pupils had remembered key points from a previous lesson, a question such as

'What can you tell me about your hands and arms if you want to get into a good handstand position?'

would be a more appropriate form of words than

'How do you do a handstand?'

Some purposes of questioning are:

★   *To make the class think*

Questions in this context are intended to make pupils think about what they are doing rather than simply following instructions with no understanding of the reason for them. For example,
'Why do you think that I asked you to put the spare mats away?'
'So that nobody could trip over them and hurt themselves.'

★   *To keep attention*

As a general principle, if a class expects to be asked questions about a demonstration or explanation then their concentration may well be improved. This category relies as much upon how the teacher handles the questioning process as on the form of the question. Pupils need to feel that they may be asked to answer. Teachers therefore ask pupils to put their hands up and wait for a number to do so before asking a child to answer.

While it is tempting for the beginning teacher to ask the first child to raise their hand to answer, in gratitude that someone is offering, such a strategy will very quickly lead the rest of the class to sit back secure in the knowledge that they will not be asked for a response. In many contexts several pupils may be asked for their answers, particularly where a number of different answers are acceptable. For example,

'Why do you think that it might be a good idea to keep your muscles tense when you try to balance?'

'Because it looks better and neater.'

'Yes, what were you going to say John?'

'Because you're less likely to fall over.'

'Yes, what did you think Ramela?'

'You can make your body shape really clear.'

★ To build up or consolidate knowledge

'What shape do your head and hands make when you do a headstand?'

'What do we mean by canon and unison?'

★ To provoke thought

'Why do you think James is managing to jump so high?'

'Can you think of a different finishing position?'

★ To test knowledge

'Which part of the beams do you have to get out first?'

'What do we call this position?' (e.g. straddle half lever).

'What is the difference between matching and mirroring?'

★ To revise earlier work

'How many different directions did you practise rolling in last week?'

'What did you have to remember to perform a good cartwheel?'

★ To check knowledge (without probing understanding)

'What do we call the part of the body which pumps blood?'

'What is the name of this vault?'

★ To test understanding

'Why does your heart rate go up when you have been running round the gym more than when you have done some press-ups?'

★ To emphasise particular teaching points

'Where does Michael place his feet to make it easy to stand up at the end of the roll?'

'When does Jaswinder push on her hands in order to do backward roll through hand-stand?'

★   *To help the pupil's observation and thereby increase understanding*

'Watch Judith. How does she use her arms to help her to keep her balance?'
'Look at Sukh's backward roll. What do you notice about the position of his hands?'

★   *To motivate and stimulate*

'Can you jump any higher?'

★   *To obtain feedback*

Questions under this heading can be used to determine ability, understanding, recall, skill level and so on. For example,
'Why were some balances easier to hold than others?'
'What factors will affect how easily you can perform a bridge?'
'How many of you feel confident about taking your weight on your hands without a supporter?'

★   *To develop reflection and assessment*

'Why do you think that Carol's performance was good?'
'How did Carl get variety into his sequence?'

★   *To give many pupils the opportunity to show knowledge or understanding*

The target of the question rather than its form is important here. Answering questions can be an important form of achievement for pupils who find physical performance difficult.

In addition to serving different purposes, questions can demand different kinds of responses from pupils. Questions which test knowledge are sometimes referred to as lower order questions while those which create knowledge are known as higher order questions (Brown 1975).

There will be many occasions when the first question does not elicit the desired answer. Indeed it may not elicit any answer at all! The temptation to provide the answer and move quickly on should be resisted. Rephrasing questions will be necessary in this situation. The following strategies may help the inexperienced:

★   **Do not be afraid to wait for an answer! This is especially important if the question is complex. A common fault among inexperienced teachers is to panic and answer the question for the class who very soon learn that there is no need for them to make any effort because if they wait the teacher will answer for them.**

★   **Word questions so that they are clear, precise and relevant to the age and ability of the class. Vague** questions such as 'What about your arms?' give pupils little chance of understanding what you mean or of constructing a reasonable answer.

★ If no one offers an answer rephrase the question *or* simplify it (see below for how to do this). It may be that pupils are not sure about your own line of thinking or it may be that they simply do not understand.

★ Order questions in a logical and meaningful way. Make the follow-up question more specific.

★ Give prompts to help pupils.

★ Encourage hesitant pupils. Ask the question before naming a pupil who is to answer. If the pupil is named first, the rest of the class have no need to think of the answer.

★ Ask a probing question to elicit further information.

★ Ask an 'either/or' question.

★ Ask an elliptical question (that is one where the pupil supplies a missing word).

Demonstrations

Demonstrations can be used to serve several different purposes.

★ *To show a technical point*

This could be how to position the head and hands in a headstand, the timing of the push in a backward roll through handstand, when to flick the hips in a headspring, and so on. It is important that what is demonstrated will actually prove helpful. For example, if what is required is a demonstration of the placing of the hands and the initial arm movement for a handspring, then this should be shown rather than a spectacular performance of the whole skill (although there is of course a place for showing the whole action). A demonstration of something which appears to be well beyond their capability can be intimidating rather than motivating if given at an inappropriate time. It may also distract from the real focus of the demonstration. For example, a through vault demonstrated with the springboard at a great distance from the box may lead to attention focusing on the distance from which the vault has been performed rather than on the technical points which are being stressed.

★ *To show something well done*

Where the teacher is working to improve quality of performance, a demonstration of this type may be used to illustrate the standard expected or the next stage of progression. For quality of performance, the level of skill is not relevant but the way in which it is performed is vital. It can therefore be a good opportunity to choose a pupil whose ability in terms of technical skill is low, but who can show good poise and finish.

★ To show a range of possibilities

This kind of demonstration can be useful when exploratory work has been carried out and can very easily involve a number of pupils of varying abilities. Alternatively, half the class could show their ideas while the other half watch to see how many ideas they can identify.

★ To compare different aspects of work

Two pupils could be asked to demonstrate headstand to handstand, one pushing up from a tucked headstand and one swinging up from a piked headstand. This could be used to compare the advantages and disadvantages of the two methods.

★ To explain a particular concept and establish its features

This might be used to explain the concepts of symmetry and asymmetry or what is meant by body tension, where the emphasis is less on how to do something and more on establishing exactly what it means.

★ To emphasise particular teaching points

This sort of demonstration may be in response to a whole class finding a particular aspect of an activity difficult, whether this be a technical point, such as keeping the elbows locked when holding the weight on the hands, or improving quality such as maintaining body tension to sustain a balance.

★ To stimulate more ambitious movement

This is particularly relevant in situations where pupils have a choice of response. It may be that one or two pupils have tried a particular answer but that the majority are producing answers which, while not inappropriate, are not challenging them. The teacher might ask a pupil to show, for example, a forward roll to straddle finish, and suggest that this is well within the capability of many other pupils.

★ To show a fault or faults

As a general rule, the teacher should demonstrate if faults are being shown rather than embarrass pupils in front of their peers.

★ To show completed work

In gymnastics or dance where composition and performance are integral to the activity, performance of completed dances or gymnastics sequences should be a feature of lessons. This sort of demonstration also provides very good opportunities for encouraging pupils to evaluate the work of others and for the teacher to make a judgement about pupil performance.

★ To show something original or otherwise outstanding

Situations sometimes arise in PE where a pupil who is particularly gifted and involved in out of school training is performing at a level well beyond that of his or her peers. A judgement has to be made by the teacher about the advisability of showing aspects of this work to the rest of the class. Sometimes pupils become very arrogant about their performance and the teacher would feel that any encouragement to 'show off' should be avoided. Others are very modest about their achievements and would be very reluctant to demonstrate in front of their peers. For others, their achievements may be a real source of self-esteem which will be enhanced by an opportunity to perform in front of their classmates.

Organising demonstrations

★ Who should demonstrate?

The teacher? Useful if a fault is being shown (with the correct version also shown) or if the teacher can exaggerate a particular point to heighten the effect.

One pupil? Useful for showing particular technical points where the teacher wants to give a specific emphasis. If pupils are chosen, are able and less able included? Boys and girls? All cultures? Has the teacher checked (a) that the pupil does not mind demonstrating (some encouragement may be needed to reassure those who do not have high self-esteem or confidence) and (b) that the pupil knows what is required?

One group of pupils? This could be one group working together to show effective co-operative work or one group working individually but demonstrating a range of ideas or skills.

Half the class? This can combine an opportunity for half the class to evaluate the work of the others with an opportunity for the performers to work with rather more space than usual.

In groups with one pupil being observed by a partner or by the rest of the group? A variation on the previous example which is useful for developing evaluation skills and for encouraging performance for others with less pressure than performing in front of the whole class.

★ Can everyone see?

It may be necessary to ask observers to move either so that their view is not obstructed or so that they have a particular view of what is being shown. For example, pupils need to see a headspring from the side in order to see the angle of the legs and the position of the hips.

★ *Do the pupils involved know what they are supposed to be doing?*

Have they been informed in advance what they are going to be asked – there is nothing more disconcerting for a beginning teacher than to be met with outright refusal having asked someone to demonstrate unexpectedly. Little better is finding that the pupil has completely forgotten what it was they were doing a few moments earlier, resulting in a totally unexpected and inappropriate demonstration. This is especially important where a pupil has produced several answers to a task.

★ *What are the observers to look for?*

Apart from the importance of pointing out key features, if that is the purpose of the demonstration, it is an opportunity for observation, analysis and evaluation by pupils which should not be missed. Initially, pupils need specific guidelines. For example, watch the other half and choose two which you think have linked their actions together particularly smoothly.

Later, pupils may be encouraged to suggest observation criteria for themselves.

Resource-based teaching

Workcards can be a powerful and effective teaching aid. They are particularly valuable when working with mixed ability groups as they offer opportunities for differentiation through enabling pupils to choose from several different challenges which can be presented on one card, or to move on to new cards as they are ready.

Workcards can be used to:

★ provide a model for pupils to match (this implies good quality pictures which show good style and correct technique);

★ provide a range of ideas to increase the range of movement vocabulary;

★ help pupils to learn specific skills;

★ offer differentiated ideas enabling pupils to answer a task at a level appropriate for them;

★ provide a progressive series of models allowing pupils to move on to a more challenging card when they are ready;

★ provide prompts to help keep pupils on task;

★ provide help for pupils when a reciprocal teaching style is used.

Factors to consider when selecting, designing or using cards.

★ How much information is presented?

★   Are the picture images clear?

★   Do they provide a good model in terms of the quality of the image presented?

★   Do they provide appropriate role models in gender and cultural terms?

★   Is the language on the card comprehensible to the pupils?

★   Does the card cater for a wide ability range or are several needed?

★   Are safety matters covered where relevant, for example the need for support or teacher assistance?

The examples at the end of this book are provided without any text so that they can be adapted to different age ranges and abilities. Suggested learning activities are given underneath each card. Text can be written on a chalkboard or whiteboard in the gymnasium or on posters placed around the walls, or can be added to the cards if they are used in plastic wallets. Greater flexibility of use can be achieved by scanning selected pictures onto a computer and generating cards for specific lessons or specific groups. If pupils have easy access to computing facilities they can produce their own workcard for use in a lesson or series of lessons. For example, the balances from cards 4, 5 and 6 could be scanned onto a computer. Pupils then choose their own specified number of balances for their personal worksheet or for their group and print them off for use in the lesson.

Many of the actions included on these cards are within the capability of younger pupils. Their relevance at Key stage 3 will depend upon the selection, by the teacher, of suitable tasks to accompany them. There are likely to be pupils in Key stage 3 classes for whom these ideas are new or who have not been able to acquire the skill in the time available at Key stage 2. Others will need to use the actions within a more advanced context if there is to be progression from earlier achievements.

For example: Page 137 shows a card with a circle roll on it. This card could be used as it is to challenge pupils whose movement vocabulary is limited. It could also be used in other ways:

★   Perform the roll on this card and follow it with a backward roll to straddle, or with a straddle lift to handstand (increasing the difficulty level);

★   use the roll shown on the card as one linking action in your sequence (composition of complex sequence). N.B. other required links could be specified (turning jump, spin on the floor, etc.);

★   work in a group of four or five and use this roll in different formations, to be performed synchronised.

Planning for gymnastics

This chapter looks at planning, first by examining the different factors which need to be taken into account when planning, whether this be at a very formal level, for example, the preparation of documentation for use within the school, or at an informal level, for example, thinking through options for a particular section of a lesson. It then offers guidance on the planning of schemes and units of work and of individual lessons.

Planning for balance

The gymnastics programme offered should be planned so that a balanced experience is provided, both within lessons and over a unit of work. Balance can be considered in relation to a number of dimensions. The questions raised below should prompt thought when planning particularly for Key stages 2 and 3. During the later phases of Key stage 3 work and during Key stage 4 non-examination courses, the need to encourage decision-making and independent learning on the part of pupils, implying some degree of choice, has to be balanced against the issues raised below. At this stage balance within the options should be considered, even if individual pupils wish to opt for a specific aspect of work.

1. The NC Order gives performance priority and this should obviously be reflected in the time allocation within a lesson. Planning and evaluation remain within the new Order and the question of the most appropriate balance between the three elements should be addressed. Opportunities for planning and evaluation can be offered through teaching styles and strategies used as well as through specific teaching episodes.

2. Is there a balanced allocation of time? Lessons are generally divided into a warm-up phase, some floorwork and some apparatus work. Ideally, lessons should be long enough to give adequate time to all three although this is not always possible, particularly where lessons are short and pupils have to walk long distances across scattered campuses. The difficulties of moving apparatus with young children in some primary schools

Jason is teaching a mixed ability Year 8 group a unit of work based on increasing range of balancing skills with a partner. The previous year he had taught several specific balances, using a practice style and then used a problem-solving approach to enable pupils to work out further balances for themselves. This year he decides to use a reciprocal teaching approach based on a number of workcards with balance ideas on them. This will enable all pupils to be involved with evaluating the work of others to a greater degree than in the previous year.

can also lead to problems. Where these difficulties arise, the time allocation over a number of weeks needs to be considered.

3. Are tasks organised so that there is balance in the demands made of the body so that actions include use of the whole body rather than giving excessive emphasis to one body part through, for example, including too much work on jumping in one lesson to the exclusion of work involving the upper body? Organising work around specific skills can result in lessons where there is excessive emphasis on one part of the body. Planning using themes can avoid this, although themes need to be carefully chosen and clearly defined. Planning to ensure that some time is given to at least three of the basic actions of gymnastics in any single lesson can provide an alternative approach. (See Chapter 2 for a categorisation of gymnastics content.)

4. Is apparatus planned to give a balanced range of activities, i.e. hanging, circling, heaving, vaulting, rolling, etc?

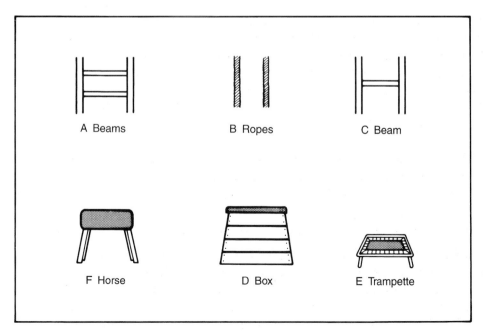

A Beams   B Ropes   C Beam

F Horse   D Box   E Trampette

Fig 7.1

> Margaret's original plan is for pupils to move around the gymnasium in a clockwise direction from A to B to C, etc. However, she realises that this will mean that pupils will be using their upper bodies on beams followed by ropes, and then repeating this. She decides instead to swap groups on opposite sides of the gymnasium, from A to F, from B to D and so on.

5. Are both fast dynamic movements and slower more sustained movements included? This question is related to the third one. Basing whole lessons around single skills or groups of skills can produce a very unbalanced experience as can work on themes, if these are not chosen carefully. For example, work based on the skill of rolling is not a satisfactory approach; neither is balance as a theme unless complemented with some contrasting activity.

6. Is apparatus organised so that apparatus changes involve a change of type of activity or so that a balanced selection of activity is possible within one apparatus arrangement? It may be that the teacher wants all pupils to experience a balanced range of activities while using a particular apparatus arrangement. It may be that the opportunities presented by the apparatus enable pupils to make a choice from a balanced selection.

> A Year 8 class is given a box, ropes, a bench and mats. They plan their own arrangement. John uses the ropes for much of his sequence because his light weight and good upper body strength means that he finds heaving activities easy and is also able to hang from his arms for long enough to show a range of body positions and to move from one rope to another with ease. Martin chooses to use the box and the bench to demonstrate a range of rolling actions, along and off the bench and over and along the box.

Both boys are thereby able to achieve success, to demonstrate competence in activities which fall within the remit of the Key stage 3 Order and to choose to work to their strengths. On another occasion the teacher might wish to encourage them to develop aspects of work with which they are less comfortable.

7. Is there a balance between practising and performing individual skills and the planning and performance of sequences? This point relates to the cyclical development of competence shown in Chapter 1. Sequence work is one example of progression but without an ongoing input of new skills/actions which can then be included, progression will be hampered. It may be that, for some pupils, the ability to develop individual performance will be through the application of existing skills rather than learning completely new ones. This will be particularly true of Key stage 3 pupils who arrive at

secondary school with a well-developed repertoire of gymnastics activities and whose performance may, as a consequence of physical changes at puberty, decline in some respects.

8. Is there a balance between teacher-directed activities and those which allow the pupils some degree of choice? While differentiation is the key to achievement by all pupils and is dealt with later in this section, it is important that a balance be struck between allowing so much choice that no progress is made and allowing so little that only a small number of pupils progress significantly.

Planning for progression

Although the new Order gives some indicators of progression from key stage to key stage, it is not as helpful as it might be in identifying indicators of progress as the pupil follows the programme of study through. These are some questions which teachers might ask.

★ Have pupils extended their personal movement repertoire?

This means increasing the range of actions and skills which the pupil can perform. It may mean acquiring new skills – a headstand, a fish roll, a flic flac. It may mean extending the repertoire through adding different dimensions – varying the leg position in a headstand or shoulder stand, ending a backward roll in many different ways, showing different body shapes in jumping to get off different pieces of apparatus. It may mean performing a previously learned skill in a different context, for example rolling forwards round a high bar rather than round a low one.

Progression from individual to pair work

Progression from floorwork to apparatus

★ *Have pupils learned new actions skills or sequences which are technically more difficult for them?*

This is about the need for personal challenge. For one pupil it may mean learning to roll backwards without going over one shoulder. For another it may mean lifting into a handstand off two feet with straight legs. For a third it may involve transferring the skill of backward roll through handstand from the floor to along a box.

★ *Has quality of performance developed?*

Initially clarity of body shape and body tension may only be achieved in held positions. Progression is shown where the same clarity and tension is present throughout the performance of first single actions, then parts of sequences and finally, through whole performances.

★ *Have pupils succeeded in integrating actions into increasingly complex sequence?*

Progression in sequence work for Key stage 3 and beyond may involve creating and sustaining quality performance through longer sequences; performing sequences using first simple and then complex apparatus; performing sequences in pairs, trios or larger groups; incorporating added dimensions such as varied pathways, different levels, changes of speed. For younger pupils it will involve linking first two and then more actions; repeating single actions to perform several linked moves; adding changes of direction and/or changes of speed. Many Key stage 2 pupils will be capable of more complex sequence work such as that outlined above.

Planning for differentiation

Differentiation is the means by which all pupils are enabled to fulfil their potential in gymnastics.

At Key stage 2, pupils will vary considerably in height, weight, strength and flexibility. During the key stage many girls will reach puberty, increasing the range of physical differences within any one class. Many Key stage 3 classes will include both boys and girls who have reached puberty alongside those who have not. By Key stage 3, pupils should have acquired a reasonable movement vocabulary including mastery of basic identifiable skills. They will be growing and maturing at different rates and will be attaining a more adult stature at different ages. This will impact differently upon their performance in gymnastics and will increase the need for differentiated experience if all are to continue to enjoy success.

Jenny is 13 and until about 18 months ago was an enthusiastic and accomplished performer in the gymnasium. At puberty she grew several inches and put on a significant amount of weight. She no longer has sufficient upper body strength to support her weight easily on her hands. The change in her centre of gravity resulting from her pattern of growth means that she finds some balances more difficult than previously. She finds her gymnastics lessons frustrating and now opts out whenever possible.

Sukh is also 13, tall and flexible but lacking in strength. He is still growing fast. He enjoys gymnastics lessons and finds the co-ordination needed to master skills easy, although good style is anathema to him and controlling his growing arms and legs can be difficult. He gets irritated with criticisms about the poor finish to his work and wants constant challenge in the form of more skills.

Karen is also 13 and has yet to reach puberty. She is slight and lightweight. She is very flexible but lacks strength although, because of her light weight and petite build, many gymnastics skills come easily. She will practise willingly and learn new skills, but resists any suggestion that she should try to improve her strength because she 'doesn't want to grow all fat and muscly'.

Mark is 13 and physically mature. He goes to weight training at the local leisure centre and has excellent upper body strength. He enjoys strength-based activities such as rope or beam work and also vaulting. He thinks floor work is very boring and regularly misses the apparatus part of the lesson because of earlier misdemeanours.

These are just four of the pupils which we might find in any Key stage 3 class. Providing them all with success and challenge is a demanding task for the teacher, but a necessary one if all are to achieve their potential. In a recent survey of Key stage 3 pupils, 23 per cent thought that they were worse than average at gymnastics, compared with only 4.5 per cent who held this view about their games performance. It is perhaps not surprising that 34% disliked gymnastics as a curriculum activity.

A number of gymnastics activities are included in the Order for Key stage 3, but there is no requirement to divide time evenly between them. At Key stage 4 there is little prescription about skills/activities to be included. There is therefore scope for taking account of interests and abilities. This is particularly important if all pupils are to remain motivated.

Given the potential material contained within the Key stage 3 programme of study, the teacher has plenty of choice and there is absolutely no need to impose the same content upon every child. Differentiation can be achieved through a number of strategies, some of which are used routinely by teachers who may not be fully aware of the differentiation potential offered.

Differentiation is often described as being by task or by outcome. Given the differences illustrated by the four pupil cameos above, differentiation by outcome, through which the same task is set to all, but with a range of acceptable outcomes, may be less successful than differentiation by task. This is with the proviso that the latter should be used for offering choice and not simply in a way which highlights ability differences. Certainly it should be possible for pupils to have some choice in the way they progress work over a series of lessons. Differentiation can also be provided through content, pace, level, access, sequence, teaching style, interest, pupil groupings and resources.

Differentiation through content

Within a specific theme, for example symmetry, or within a specific action category, for example, rolling, different pupils may work within a different content emphasis. Within the context of symmetry, some pupils might be attempting a variety of symmetrical balances while others might be trying different flight activities with a symmetrical body position. In the context of rolling, some pupils might be learning a circle roll, others a fish roll, others a backward roll to straddle.

Differentiation through content: inverted balances on floor

Differentiation through content: inverted balances on apparatus

Differentiation through pace

In the context of physical education, the pace at which pupils perform or practise may well be a function of physical fitness and capacity as well as cognition or concentration. For example, some pupils may only be able to sustain activities which involve taking the weight on the hands for a short period, whereas others may be able to sustain such activity for a long time. Use of workcards, which enable pupils to move on when they have finished the tasks on one, can provide opportunities for all to be challenged at their own pace.

Differentiation through level

This may be related to differentiation through pace. Where workcards are progressive, perhaps through setting increasingly difficult or more complex tasks, pupils may be working at different levels as well as at a pace appropriate to their abilities. For example, some pupils might be working on a tucked headstand, others on a headstand with a straight body, others on a headstand with straight legs throughout, others on a headstand push up to handstand.

Differentiation through access

In order for all pupils to have access to particular activities it may be necessary to modify the presentation for some. A number of pupils will only be able to attempt some skills if physical support is available, either for safety or for confidence. Some pupils may have very particular needs, such as brightly coloured benches to enable visually impaired pupils to use them safely.

Differentiation through sequence

While some skills have certain prerequisites, for example the ability to support the weight on the hands before attempting to learn a handspring, others may be attempted in any order. Pupils can be given apparatus choices so that they choose whether to develop skills of balancing using a beam, or hanging and swinging using ropes, or flight using a box.

Differentiation through teaching style

Good teachers use a range of teaching styles and strategies without necessarily thinking of their use as a form of differentiation. Pupils, however, often respond differently to different approaches and some will match individual's preferred learning styles better than others.

Differentiation through level:

handstand on floor with apparatus as support

handstand on floor unsupported

handstand on low apparatus with help

handstand on high apparatus

Differentiation through interest

Particularly at Key stage 4 or the latter stages of Key stage 3, lesson content may be negotiated to reflect pupil interests. This form of differentiation could be through allowing pupils to choose their own apparatus, or whether to develop apparatus or floor skills. It could be through giving choices of activity, such as the opportunity to choose between different floor skills, elementary or advanced. One pupil might want to learn a flic flac, another a double foot lift into handstand, another a front somersault, another a free cartwheel.

Differentiation through pupil groupings

Pupils often work in friendship groups, however this form of grouping may not be the most appropriate for all gymnastics work. Grouping according to a combination of size and ability may enable apparatus to be put at different heights to accommodate all. Some pair work may be facilitated by pairings of similar size and weight if both partners are to take the role of performer and supporter. Alternatively, pairs of contrasting weight may enable one to play the supporting role while the other is the performer. In other contexts, allowing pupils to choose whether they work alone, in pairs or in larger groups may also accommodate individual differences and preferences.

Differentiation through resources

This form of differentiation can be provided in various ways. If some pupils use work-cards and peer group teaching, the teacher may be freed to give time to others who need extra teacher attention. Choice of whether to use apparatus, or of which apparatus to use, as mentioned earlier, is another example of differentiation through resources. In some cases adapted resources may be needed for some pupils, such as lower apparatus for a pupil who suffers from epilepsy.

Planning for equal opportunities

Where schools make decisions to implement a part of their physical education programme in mixed-sex groupings, gymnastics is often one of the activities chosen as particularly suitable for this work. In some schools, mixed-sex work is introduced from necessity because of staffing imbalances or changes in proportions of male and female pupils. I am going to assume here that decisions to teach mixed-sex gymnastics will have been made from choice and outline some of the questions which need to be answered before the decision to work in this way is made and in order to give maximum chance of its successful implementation.

★ *Is current physical education clothing appropriate for mixed-sex work?*

In both dance and gymnastics, where presentation and performance are key elements of the activities, it is particularly important that pupils feel comfortable and at ease with their appearance. Embarrassment and self-consciousness can only have an adverse effect on both learning and performance.

'I do not think that we should be made to do things in front of boys when we are only wearing gym knickers and T-shirts.' (13-year-old girl)

'Our religion does not allow us to show our legs and bodies in public but our teachers try to make us.' (15-year-old girl)

'The girls are allowed to wear tracksuits but we have to wear our shorts – it's not fair.' (13-year-old boy)

Where cultural factors require that the body should be covered in the presence of unrelated males, lack of sensitivity may result in pupils absenting themselves from physical education altogether.

It is easy to underestimate the negative effect that the wearing of inappropriate clothing can have on adolescent pupils. It is also all too easy to find examples of pupils being expected to wear clothing which contradicts all that they have been taught about safe exercise and the importance of warm muscles. Pupils should be able to wear tracksuits, sweaters/sweatshirts, leggings or tights or other appropriate clothing to keep them warm and to reinforce the message given to them elsewhere about warming up and keeping warm during exercise.

★ *Will any cultural issues affect the success or otherwise of mixed-sex work?*

This is a question for each individual school to address. While respect for different cultures should be a cornerstone of every school's equal opportunities policy, the practical implications of this will vary from one locality to another and will depend upon community links, communication and trust. In some communities, mixed-sex work may be seen as quite acceptable provided that certain rules about clothing are upheld. In others, there may be considerable resistance to mixed-sex work and pupil interests may be better served by single-sex teaching.

★ *What is the attitude of the staff in the department to working with pupils of the opposite sex?*

It is only comparatively recently that initial teacher education courses in physical education have become mixed sex and, consequently, a majority of physical education teachers trained in a single-sex environment where issues about teaching groups of the opposite sex or of both sexes did not arise. There may well be considerable hostility to mixed-sex teaching and, in this situation, it may be that pupil learning will suffer more if the policy is forced through against the will of some staff than it would be remaining in single-sex groups. However, this does not mean that it is acceptable to offer the single-sex groups totally different curricula.

★ *Have boys and girls followed a common curriculum in the past, albeit in single-sex groups?*

The history of school gymnastics has been one of markedly differing programmes for male and female pupils in many schools, reflecting the male vaulting and agilities tradition compared with the theme-based educational gymnastics approach favoured by many female departments. The introduction of successful mixed-sex work will require discussion and compromise rather than the imposition of one tradition over another.

★ *How is content to be selected to maximise every individual's opportunities for success and achievement?*

Early socialisation processes continue to ensure that boys and girls acquire, at an early age, clear conceptions of appropriate 'male' and 'female' abilities and interests in the context of physical activities. Gymnastics is often seen as a 'female' activity, characterised by the ability to perform cartwheels and the splits and requiring considerable flexibility. While it would be good to think that the primary school would overcome such ill-informed stereotyping and that pupils would transfer to secondary school and to Key stage 3 aware of the potential of gymnastics to develop a variety of physical capacities, the reality is that Key stage 3 pupils are likely to continue to arrive with inaccurate assumptions about physical qualities needed to be successful in the gymnasium.

Planning for special educational needs

Numbers of pupils with special educational needs in mainstream schools have increased significantly in recent years. There are also many pupils in schools who may well have special educational needs in the context of physical education. There is not room here to consider all the aspects of special educational needs provision. General principles can, however, be identified.

★ Because gymnastics takes place indoors and can be undertaken as an individual activity, accommodating pupils with special educational needs should be more straightforward than in some other activity areas.

★ If the focus is on what the child *can* do rather than on his or her limitations, then relevant and related activities can usually be found. For example, the child who has limited use of the legs may well be able to substitute hanging from the hands rather than balancing on them, or sliding rather than stepping as a form of travel.

★ Because learning in the gymnastics context can be less language dependent than other areas and subjects, pupils with learning difficulties may well be able to achieve good levels of performance provided other means of communication such as visual models are used. In addition, there are opportunities to foster language development within gymnastics lessons.

The other group of pupils with special educational needs is the gifted. In gymnastics, where high levels of performance are reached by young children, particularly girls, there may well be pupils who attend gymnastics clubs outside school and who have a range of skills well beyond that which could be expected within a curriculum context. For very high-level performers, who are already training for long hours, participation in curriculum work is likely to be of little benefit. Such pupils need special treatment which could involve allowing them to spend physical education time on homework to compensate for the time given to evening or early morning training. For others, participation can be an opportunity to apply advanced skills in a different situation or to develop aspects of planning and evaluation in the context of helping others.

Planning units of work

Units of work should include:

★ information about intended learning outcomes, that is, what pupils can be expected to do, know and understand by the end of the unit;

★ how these outcomes relate to the National Curriculum Order, both activity specific and the relevant EKSD (End of Key Stage Description);

★ the assessment indicators which will be used to check pupils' learning;

★ what learning activities will be included in order to achieve the intended outcomes;

★ what resources will be needed;

★ an indication of learning activities/outcomes for each of the unit lessons.

An example is given below of a unit of work for a Year 9 group.

Example of a unit of work – Year 9 (6 lessons)

Learning outcomes
By the end of this unit pupils will have:

★ experienced a variety of partner or trio balances;

★ refined their performance of several chosen partner or trio balances;

★ used partner or trio balances in a sequence;

★ used partner or trio balances with apparatus as single skills or within sequences (extension work for more able/confident).

NC elements addressed: refine and increase range of balancing skills including moving fluently into and out of balance; develop, refine and evaluate a series of actions with contact with others; refine series of actions in increasingly complex sequences.

Relevant EKSD: plan/compose more complex sequences of movements; adapt and refine skills.

Assessment indicators: can perform a number of partner balances; can use partner balances in a complex sequence; can apply partner balances to apparatus.

In order to achieve these learning outcomes pupils will:

★ learn to perform two set partner balances (performance);

★ select further balances from workcards or create their own (performance and planning);

★ practise and refine their performance of chosen balances (performance and evaluation);

★ further refine performance to include moving into and out of balances (plan, perform, evaluate);

★ link two partner or trio balances together (plan, perform);

★ add further balances and links to produce longer/more complex sequences;

★ refine their performance;

★ add apparatus (if desired).

Resources needed:

★ mats;

★ selection of apparatus for later lessons;

★ workcards (see page 143).

Learning outcomes for individual lessons (adapted according to length of lesson, pupil progress, pupil abilities):

Introducing partner/trio balances

★ pupils will appreciate safety factors:
  – sensible behaviour;
  – safe body parts for supporting weight; unsafe body parts;
  – implications of relative size and strength for who takes the role of base/supporter;

★ learn to perform two balances chosen by teacher;

★ pupils choose and perform further balances from workcard with limited choice.

Further work on partner balances – increased range and refined performance

★ recap of points from last lesson, especially safety;

★ pupils work from workcards with new ones available to increase choice and provide further challenge – aim to extend repertoire of balances.

Refining performance including moving into and out of balance

★ pupils select from the balances they have learnt and refine performance of their chosen ones;

★ pupils begin apart from each other and move together and into balance;

★ pupils move out of balance and away from each other;

★ pupils identify other balances which they can perform and which they can move into and out of in good style (NB this may mean that some balances are discarded because pupils are unable to move into them without help or with any degree of control).

Development of sequence work

★ pupils will link first two balances and then others (length of sequence dependent upon ability);

★ pupils will decide whether one will take the role of base throughout or whether they will alternate;

★ pupils will select appropriate linking actions (dimensions such as direction, pathway, speed may need revision);

★ pupils will revise/practise some chosen linking actions.

Further development and refinement of sequence work

★ introduction of apparatus if desired;

★ pupils will choose either to work on a floor sequence or to adapt balances to apparatus.

Practice and performance of sequences on floor or on apparatus

★ pupils will identify criteria for good quality performance and will judge each other's performance against their chosen criteria;

★ pupils will perform their work for another pair/half class/whole class to assess.

Planning individual lessons

The level of detail with which lessons are planned will vary depending upon the experience and expertise of the teacher. For the student teacher or for one who lacks confidence in their ability to teach in this area, individual lesson plans should include:

★ intended learning outcomes;

★ resources needed;

★ assessment indicators;

★ details of learning activities and the learning focus;

★ an indication of timing within the lesson.

An example of a Year 9 lesson from the unit of work previously outlined is given below.

Year 9

Learning outcomes: Refined performance of contact partner balances.
    Smooth movement into and out of balance.
    Sensible judgements about own performance and use of these
      to improve their work.

Resources: Workcards.
    Mats.
    Chalkboard or whiteboard.

Assessment indicators: More controlled and refined performance of balance.
    Smooth movement into and out of balance.
    Independent improvement of performance.

One member of group leads warm up

PHASE	LEARNING ACTIVITY	LEARNING FOCUS
Warm up (5 mins)	Pupils in groups of four or five. One member of group leads warm up for that group.	Independent warm up. Application of principles of healthy exercise.
Floorwork (10 mins)	Get out mats. Give out partner balance sheets (pupils choose sheet).	
	Revise some of the balances done in earlier lessons or try new ones.	Refinement of performance and/or increased range.
Development – linking actions (10 mins)	Choose one balance and find a way of moving smoothly into that balance starting some distance away from partner.	Development of basis for complex sequence work. Applications of previously learned skills to make effective links.
(10 mins)	Ask pupils what criteria they would use to assess themselves. Write performance indicators on board.	Development of evaluation skills.
	Watch several pairs performing and analyse their performance.	Development of evaluation skills.
(15 mins)	Choose a balance and move smoothly out of it.	Basis for sequence work.
	Choose other balances and move into and out of them. After moving out of one balance, move back together to perform a second balance.	Using planning skills. Development of sequence work.
	One pair watch another and comment on their performance.	Development of evaluation skills.
	Put mats away.	
Concluding activity	Choose an individual balance and practise it.	

CHAPTER 8

Apparatus work

Work with apparatus both large and small provides opportunity for new challenges and for the adaptation and extension of work done on the floor. Large apparatus both presents challenge and excitement to most pupils and opens up possibilities for achievement in activities which some have found difficult on the floor. The National Curriculum Order makes specific reference to apparatus work but also includes requirements to 'extend and refine . . .' where part of extending the range of possibilities would be to adapt the action to various apparatus pieces.

Because pupils usually enjoy using apparatus and find it exciting, it is easy for this part of the lesson to degenerate into activity without any teaching or progression. Once an initial period of exploration is over, then tasks must be set so that the work undertaken is purposeful and of high quality. Few schools have sufficient apparatus to enable the teacher to give groups identical apparatus arrangements (if this is desired). Consequently even if the same task is set for the whole class, different apparatus will produce variations in response. This makes considerable demands on the teacher if good quality movement is to be achieved and progress maintained. In setting out, using and putting away apparatus, organisation is clearly essential. The selection and arrangement of the apparatus needs careful consideration so that the challenges offered are appropriate to the work being demanded. This chapter looks at the organisation and handling of apparatus, at safety factors in the use of apparatus, and at the selection and arrangement of suitable pieces of apparatus for particular kinds of work.

Organisation and handling of apparatus

It is the responsibility of the teacher to ensure that a safe gymnastics environment is provided. The following points should be observed.

1 Discipline and control are essential in all lessons.

2 Pupils must be aware of safety rules and the teacher must constantly enforce these.

3 Apparatus should only be used by pupils after it has been checked by the teacher. If a class uses a piece of apparatus which is insecure or unsafe, the teacher may be held responsible for any accident which may occur.

4 Once working time on apparatus is over, pupils should not be allowed to continue to use it.

5 The floor of the gymnasium should be smooth and non-slip. (This of course applies equally to floorwork.)

6 Apparatus should be checked and repaired regularly by a suitable specialist company. Broken or insecure equipment should not be used.

7 Apparatus should be positioned to avoid collisions between pupils and walls, partitions, low ceilings and so on. There must also be sufficient space for pupils to work without colliding with each other.

8 Where potentially hazardous skills are attempted, the teacher should be prepared to stand in and support. Pupils should not be allowed to attempt such skills without the necessary support. The teacher should not attempt to teach these skills unless they have been trained to do so.

9 Pupils should be suitably dressed for the activity.

10 The teacher should familiarise him or herself with the operation of the various pieces of equipment available before attempting the use of such equipment in a lesson.

It should also be noted that a class whose handling of apparatus appears to be good when observed being taught by an experienced teacher will not automatically continue such good habits. Authority needs to be established and respect gained before a class can be safely left to handle large apparatus with a minimum of supervision.

Teachers are often told that classes have been taught to get out all or some of the large apparatus available. It is as well to bear in mind that this does not mean that they will remember or that they will have used such apparatus in the immediate past. Constant reminders and revision are necessary, particularly with groups whose attention span is limited or whose concentration is poor.

When in doubt err on the side of safety.

A suggested procedure for getting out apparatus

Ultimately, the way in which a class is trained to handle apparatus will depend on the preference of the individual teacher, particularly in such matters as the extent to which they wish to give the pupils responsibility and choice of action. The procedure suggested here would be a suitable starting point with an unknown class.

Before the lesson

Arrange apparatus so that later organisation is made as straightforward as possible. It may be possible to go into the gymnasium before the start of the lesson, in which case the room can be prepared in advance. If not, try to set the class working while you quickly arrange the room, i.e:

a) Put horses, boxes, etc., at the side of the gymnasium close to where they will eventually be needed, or at least get them out of the apparatus store and put them at the end of the gym;

b) Put mats ready in an accessible place. Ideally have a pile of mats in each corner of the gym so that there is no need for queuing.

Pupils can be taught to carry out both of these tasks on first entering the gym.

During the lesson

Work to a routine.
Always stress safety before speed.
When getting large apparatus out, follow the routine below.

a) Put away any mats or small apparatus which will not be needed. Move the rest close to its new position if necessary.

b) Adjust the size of working groups if necessary and sit class down in groups where their apparatus is to go.

c) Tell each group exactly what they are to get out and how it is to be arranged. Do not let any group begin to get apparatus out until you have told everyone what they are to have and made sure that they have heard by questioning them. With younger pupils, either give specific tasks to each member of the group (e.g. Jane and Marsha get out the bars, Mark and Sukhbir fetch a bench, Sarah and Jaswinder rearrange the mats) or give them time to sort out who is going to do what.

    The use of apparatus cards with a clearly drawn diagram can save considerable time and are highly recommended. Alternatively a plan can be drawn on a chalkboard. Some explanation will still be needed.

d) Remind groups how to handle their particular apparatus, i.e. the order in which the bars come out and the need to ensure that the upright is bolted into the floor before getting the bars down. Remind groups how many people are needed to get horses and boxes out and how the wheels operate, or, if there are no wheels, how to carry them. Remind groups how wallbars or window ladders come out and how to make sure that they are bolted into place before attaching any other apparatus, for example, inclined benches, to them. Some of these reminders can be put onto apparatus cards but remember that alternatives will be needed for pupils with reading difficulties.

e) Tell groups to sit down beside their apparatus as soon as it is out. You can then either wait until everyone is ready or allow individual groups to begin working once you have checked their apparatus. It is important to be strict about forbidding pupils to use apparatus before they have been given permission.

f) While groups are getting out apparatus, position yourself so that you can see everyone. If you have to help with an awkward piece of apparatus, try to avoid turning your back on the rest of the class. Do not disappear into the apparatus store – if you stand at the entrance you can direct operations both in the store and in the main body of the gymnasium.

g) Check that each piece of apparatus is:

★ safely positioned in relation to other groups and the walls;
★ correctly set up.

h) Set the task and start the class working.

At the end of the lesson

a) Sit the class down on the floor while you remind them of the order in which things are to be put away.

b) All benches, etc., must be unhooked from large apparatus first. Mats and benches will have to be moved or put away before bars or window ladder can be put away. Remind again of the way in which apparatus is to be handled, also of the number of people needed to carry various pieces of equipment. The best way of reminding the class is to ask them to tell you.

c) Stress safety before speed.

d) Set a task to be done on the floor once the apparatus is away and see that it is done.

e) Position yourself so that you can see everyone while the apparatus is being moved. Be ready to give a hand with awkward or stiff pieces, for example, putting boxes onto their wheels, unbolting window ladders.

When an apparatus arrangement is being used for several lessons, it will save time if groups get out apparatus with which they are familiar. Either they get out the apparatus they put away at the end of the previous lesson and then move on to a new arrangement or move on at the end of the lesson and put away the apparatus you want them to get out the next time.

Safety on apparatus

Reference should be made to the BAALPE publication *Safe Practice in Physical Education* for advice on safety matters related to gymnastics, particularly with respect to maintenance and use of apparatus.

Selection of suitable apparatus arrangements

1 Apparatus must have sufficient space for safety and to enable the task set to be completed.

Make sure that members of the group can return to their starting point without hindering the work of others. It may be necessary to specify the route to be used.

If two groups are working in opposite directions they will be able to wait for turns together, rather than one group waiting where another is trying to work.

There will be many situations where there should be no need to form queues for turns.

In certain situations it may be advisable to group classes according to height and provide apparatus of different heights to cater for the different groups. Graded apparatus may alternatively be needed so that pupils may work progressively.

2 Apparatus should be selected so that the theme or topic which is being used may be explored and developed successfully on the apparatus as well as on the floor. For example, if a class is working on *levels*, then the apparatus should be selected so that changes of level are encouraged. Therefore arrangements where everything is at the same height would not be suitable for this theme. Similarly, where different pathways are desired, the arrangement must allow for a variety of approaches and this has implications for the amount of room that will be needed around each arrangement.

It is rarely possible, and would in any case not always be desirable, to have the whole class working on identical apparatus arrangements, all attempting the same task. Some thought therefore needs to be given to the combinations of type of apparatus and task, so that logical progress may be made. Although many schools may only have a single piece of apparatus, most have similar *types* of apparatus which may be used for an identical task or with only very minor modification to allow for slight differences in the possibilities offered. For example, a school may only have one box, but by using horse, buck, and other padded flat surfaces (possibly improvised such as using a bench with mat or mattress on top) will be able to provide several arrangements suitable for vaulting type or balancing activities. Some examples of how apparatus may be planned are given below.

a) All groups use the same apparatus or type of apparatus. All groups are set the same task. Only possible when numbers are small or a lot of apparatus is available. For example, six boxes/flat surfaces lengthways with mats. Get onto, along and off, over apparatus using hands and feet only.

b) Each group is given different apparatus. All groups are set the same task. For example:

★ double bars + mats;
★ horse + mats;
★ box + mats;
★ wall bars + inclined bench + mats;
★ window ladders + mats.

Get over, across, through or along apparatus using hands and feet only.

c) Two types of apparatus given. Two contrasting tasks set. For example:

★ four groups on wallbars + inclined bench + mats;
★ four groups on boxes/horses + mats.

Use wallbars, benches and mats to make up a sequence showing different types of weight transference without flight. Get on/off or over boxes, etc., showing flight onto or off the apparatus.

d) Using eight groups, four types of apparatus are given. All groups are set the same task. For example:

★ two groups on boxes + mats;
★ two groups on double bars + mats;
★ two groups on window ladders + mats;
★ two groups on horses + mats.

Use your apparatus to show twisting movements.

e) Apparatus as in d). Each group is set a different task. For example, get onto boxes and horses using hands and feet only and get off using a turning jump. Turn round the bar or window ladder either forwards or backwards and then twist to come down to the floor.

f) Each group given different apparatus. Each group set a different task. For example:

★ bars, inclined bench, mats;
★ window ladder, box, bench, mats;
★ horse, bench leading to it and away, mats;
★ ropes, buck, mats;
★ single bar, bench parallel to it, mats;
★ wallbars, ropes, bench, mats.

Use bars and inclined bench to make up a sequence as a group showing individual and assisted flight. Use window ladder arrangements to make up a group sequence showing twisting, turning and flight. Use horse arrangement to practise individual vaulting skills. Use ropes and buck to make up a sequence in twos showing matching actions. Use bar and benches to work out a synchronised group sequence. Use wallbars and ropes to make up individual sequences showing flight and balance.

Balance using apparatus with weight at least partially on hands

In the above examples an initial task only is given. Where different apparatus arrangements are used, the same task cannot be given to the whole class indefinitely, because
it needs to be wide enough to cater for the differing demands made by the various
pieces of apparatus. As a result, the task will eventually be too wide for purposeful work
and progression to be achieved. In order to increase skill and to produce high-quality
work, the pupils need to work within a tightly constrained situation with an element
of choice to cater for individual differences and to encourage a personal response if this
is desired.

CHAPTER 9

Assessing work in gymnastics

What do we mean by assessment?

Put simply, assessment involves making judgements about pupils for specific purposes. Different approaches to assessment reflect different philosophies and will have different effects upon those involved in their use. Within gymnastics, teachers use a variety of approaches during their day-to-day work.

Approaches to assessment

Ipsative assessment

This means that performance is compared with one's own previous best and is assessed irrespective of what others have achieved. It indicates whether a pupil has improved or not and how much progress has been made. For example, last year Sukh was able to roll backwards but landed on his knees and went crooked. Now he is able to maintain a straight line and to regain his feet. Judith could kick up to a handstand from a single take off last term. Now she is able to use a double take off successfully. Using this approach it would be possible for everyone in a class to achieve a high grade because all had improved on an earlier performance, even if the level of the better performance was very low.

Ipsative assessment will record progression but not necessarily absolute performance levels. It should motivate those whose achievement will never match that of their peers.

Norm-referenced assessment

This means assessing a pupil's performance in relation to standards achieved within a given group. It enables comparison of one pupil against another. For example, two thirds of Year 8 are able to perform a backward roll to straddle and to hold a tucked headstand. Those unable to perform these actions are below average. By definition, some members of any group will record a poor performance. Selection of representative teams involves making norm-referenced judgements since choosing the best players involves making comparisons with the cohort available rather than against absolute standards. An U13 team could well be able to beat an U14 team because average standards of play are higher in Year 8 than in Year 9. This approach to assessment is less obviously applicable in the gymnastics context than others, although teachers will compare pupil performance to their view of what constitutes 'average'.

Norm-referenced assessment will record performance level but not progression. It will motivate the successful but may well have the opposite effect on others.

Criterion-referenced assessment

This means assessing an individual's performance against previously set out criteria. It assesses the extent to which agreed goals have been achieved. For example, the criterion might be that pupils should be able to perform a sequence including specified movements – a turning jump, a backward roll, a handstand. James includes all of these together with an arab spring and a forward roll to straddle. He therefore meets the criteria comfortably. Jaswinder is able to include a turning jump and a backward roll but substitutes a headstand for a handstand, which she finds difficult. She does not meet the criteria. The criteria could be that the sequence should include a turning jump, a backward roll and a balance where part of the weight is supported on the hands. Jaswinder would now meet the criteria. There are no predetermined limits on the number of pupils who can achieve the expected standard. Governing body awards use criterion-referenced assessment in that all pupils who can demonstrate the requisite skills may gain the award. It should motivate pupils by giving them clear goals for which to aim, provided that these are realistic, and can provide a summative judgement about performance.

Formative assessment

Formative assessment takes place during the learning process. Its purpose is to describe progress and to identify pupils' needs. It may utilise a whole range of assessment strategies in order to inform teaching decisions on a day-to-day basis. 'Many pupils were not being given a sufficiently clear idea of their progress or an indication of how they might improve the quality of their work.' (DES 1992)

Summative assessment

This takes place at the end of a unit of work and aims to measure and record attainment through summarising pupil achievement rather than to influence teaching. The

Martin is practising a backward roll and trying to go through to a handstand position. His body tension and style are excellent and he has considerable upper body strength but he is pushing too late from his hands and is not getting into an inverted position on his hands. Which of these two examples of feedback would be more effective in the context of formative assessment?

'Martin, you are not getting into a handstand at all.'

'Martin your actual roll and the strength of the push are excellent. If you can push from your hands earlier and try to aim your feet towards the ceiling you will probably manage the handstand.'

information may be provided for pupils, for parents, for governors, for employers, for other teachers or for any combination of these. Harlen (1991) describes two approaches to summative assessment as summing up and checking up. The former is essentially a summary of the formative assessments which will have been made over a period of time while the latter collects 'new' information about a pupil usually through some form of test.

Recording achievement is one way of summarising aspects of formative assessment in that it gives the pupil the opportunity to describe their current level of achievement over a range of indicators and to set themselves targets for the immediate future.

If summative assessment is based only on the formative assessments made then the outcome depends upon the pupil having had the opportunity to show skill, knowledge or understanding. It may also be based upon information which is now out of date.

Amanda's record shows that she can perform simple skills and can link them to make a short sequence following a half term unit in the autumn term. However, by the end of the year Amanda can actually perform a greatly increased range of skills and more complex sequences because she has been attending gymnastics club since Christmas.

This can be particularly true of physical education where pupils may well be involved in activities out of school and be given opportunities to extend their learning and their performance independently of the school.

'Checking up' in the form of tests does give all pupils a chance to show what they have learned. On the other hand they take a long time and decisions have to be taken as to whether the loss of teaching time can be justified for the additional information provided.

What does assessment involve?

Teachers and students need to:

★ *know what they want pupils to learn*

This may appear to be stating the obvious, but, for the beginning teacher, clarifying exactly what the focus for a lesson is can be an important part of the planning process. For example, asking pupils to work in groups of three or four to plan a gymnastics sequence could have a number of outcomes:

a) pupils could demonstrate high levels of individual skill with few common elements to the work;

b) pupils could assist each other, for example, by including partner balances or supported activities;

c) pupils could plan a piece of work in which each individual included two movements of his/her choice but which also included some common actions;

d) one pupil could 'choreograph' the work of all the others.

The judgements which the teacher made about the work would depend upon whether his/her learning priorities were:

★ individual performance in which the first outcome would be acceptable;

★ effective group interaction in which the second outcome would be acceptable;

★ appreciation of the strengths and limitations of group members in which case the third outcome would be acceptable;

★ group roles including leadership in which case the fourth outcome would be acceptable.

If the teacher is not clear about the possible outcomes from a particular learning activity then they will find it difficult to make an assessment of what has been learned. If his/her anticipated outcome is not achieved it is easy to jump to the conclusion that the pupils have not learned anything when, in fact, they have learned something equally worthwhile.

★ *recognise what is and what is not assessable*

This is particularly important at a time when physical education, in common with other subjects, is expected to be accountable. Teachers therefore need to be able to demonstrate to senior staff, to parents and to governors that their programmes have clearly articulated aims and that they achieve these. For example, the aim of preparing pupils to lead active adult lives is very laudable and few would disagree with its inclusion. However, its achievement is not assessable in the short term. Other objectives are therefore needed against which the programme may be judged.

★ *have analysed the development and progression involved in the learning process*

Without understanding the various learning stages involved in specific activities, the teacher will find it difficult to judge whether or not a pupil is making progress towards achieving success or whether they need guidance to change aspects of technique or strategy. For example, moving from successful performance of a skill in isolation to synchronising it with a partner or incorporating it into a sequence constitutes progression. If a pupil cannot perform a skill in the new situation they may need help to adapt the context slightly as an interim stage.

★ *provide opportunities for pupils to learn*

If pupils are to learn new skills and tactics they need time to practise and refine their actions. This means that they need to learn to practise effectively without constant intervention by the teacher.

★ *identify sources of evidence of attainment/achievement*

Teacher observation will provide a lot of evidence of performance but not necessarily of understanding or knowledge. Other sources of evidence are important particularly for pupils who find performance difficult but may nevertheless have a very good understanding of the factors involved in the successful performance of a gymnastics skill. Assessment of the planning and evaluation component of the physical education attainment target requires more than simply observing pupils perform. Some other sources of evidence might be:

★ pupil self-assessments or peer assessments;

★ written materials, e.g. notes, diaries, workbooks, written descriptions or plans, diagrams of patterns of play, sequences;

★ group discussions;

★ answers to questions;

★ pupil explanations or descriptions.

The teacher needs to be able to provide constructive feedback to pupils at all times, and to make an assessment of attainment or progress from time to time. Evidence of achievement is often easy to see, for example the pupil succeeds in swimming a width non-stop, or in standing up at the end of a forward roll without using hands. At other times it is less obvious, for example, the pupil has a very good understanding of what is needed to perform a forward roll but is not yet able to put all of this knowledge into practice. In these cases the teacher needs to identify what has been achieved and what needs further work.

Principles of effective assessment

★   assessment should be on the basis of agreed criteria which are known to pupils;

★   criteria used should reflect expected educational development and progression;

★   assessment should not interfere with normal teaching and learning;

★   assessment should be related to a manageable recording and reporting system.

It is very important that pupils know what is being assessed at the beginning of any unit of work. Assessment criteria can be written out and put on notice boards or on the wall in the changing room of the gymnasium. Pupils should be able to refer to them throughout the unit of work.

Developing assessment criteria

1   Establish general assessment criteria related to each EKSD.

2   Establish activity specific criteria related to the general criteria. That is, what do you want the pupils to learn within the activity area, and in what order?

3   Relate activity specific assessment criteria/learning outcomes to the units of work within your programme.

4   Make decisions about assessment, recording and reporting strategies.

Remember that the end result needs to be manageable! There is no requirement to assess everything to assess at the end of every block of work.

Page 117 shows an example of a recording sheet for a unit of work at Key stage 3. It is designed to be used in conjunction with the class register, thereby avoiding a separate and additional task for the teacher.

Page 118 shows an example of a simple form for recording progress and achievement during a key stage (adapted from Spackman 1995).

KS3, YR 8 GYMNASTICS	Assessment aim A Extend/refine skills			Assessment aim B Compose sequence			Assessment aim C Work with partner			
KEY 1. Working towards 2. Can do 3. Can do well	Assessment criteria Weight on hands using apparatus			Assessment criteria Plan sequence including specified actions, variety, contrast			Assessment criteria Share roles, assess each other's work			
		A			B			C		
NAME	Attendance	1	2	3	1	2	3	1	2	3	Kit
Andrews Mark											
Anthony Clare											
Bloom Adrienne											
Boxer Mandy											
Brixton John											
Bromfield Solomon											
Christie Sally											
Clover Nathan											
Drew Neil											
Durbin Geoffrey											
Eccles Kelly											
Exbury Tim											
Glover Matthew											
Harvey Winston											
Jones Evan											
Kaur Ravinder											
Khan Naseem											
Matthews Jane											
Singh Sukbir											
Thomas Jenny											
Troughton Hugh											
Wall Jack											
Williams Emily											
Wolley Jo											

RECORD OF PROGRESS AND ATTAINMENT – PHYSICAL EDUCATION	
NAME:	**KEY STAGE:**
UNIT OF WORK	**EVIDENCE**
Gymnastics – planning and composition; refining and increasing range of actions; cope with success and limitations	
Games (winter) – devising strategies and tactics; development of variety of techniques; appreciate strengths and limitations	
Games (summer) – devising strategies and tactics; development of variety of techniques; appreciate strengths and limitations	
Dance – composition, performance with control and sensitivity to style	
Swimming – two recognised strokes; apply rescue and resuscitation techniques	
HRE – preparation and recovery from exercise; effects of exercise on body systems; role of exercise in maintenance of health	

Recording and reporting

Schools will have their own systems for the formal recording of progress and reporting to parents. What is equally important, if not more so, is the day-to-day informal reporting to pupils on their progress. This is particularly true of a subject like physical education where the outcome of the pupils' efforts are immediately visible and public.

'Well done, Sarah, that's the best handstand I've seen and it's really improved since last week.'

'Your sequence work has really improved, Mark. If you concentrate on firm muscles and straight

legs all the way through and practise your headstand to keep your balance, it will look even better.'

I'm disappointed in your work today, Anne. You can do very good work but you won't improve unless you concentrate on practising the things you find difficult.'

Examples have already been given of recording sheets for use by teachers. Involvement of pupils in self- or peer-assessment is also desirable. It is entirely consistent with the expectation that pupils should be involved in evaluating their own and others' performance. Pupil assessments can also provide evidence for the teacher of the planning and evaluating component of the programme of study. The example below is of a pupil's recording sheet developed from workcard number 37.

Task

With a partner, choose four of the balances below and link them to form a sequence which includes rolls, spins, jumps and stepping actions. Perform the sequence so that you synchronise with your partner and so that you show changes of speed.

Observer/assessor

1. Tick the balances which are used.
    Use the back of this sheet to answer the rest of the questions.
2. Write down the actions used as links.
3. Did you see changes of speed? Where?

A recording/assessment sheet for more advanced work is given below.

PERFORMERS

Plan and perform a sequence which uses all the apparatus to include the following:

★ One group balance involving everyone;

★ The involvement of everyone in one partner balance;

★ Each member of the group using all pieces of apparatus;

★ Variety and contrast.

ASSESSORS

Decide on the criteria you are going to use to judge the performance and write them on the back of the sheet in the left-hand column.

Look at each of the criteria you have chosen and discuss what you will expect to see in a very good, a satisfactory and a poor performance.

Watch the sequence at least twice.

Come to an agreement about your judgements and about why you have made them and write your comments in the right-hand column.

10

Gymnastics' contribution to broader curriculum areas

General physical education requirements related to gymnastics

General requirements apply across all activity areas and throughout the key stages. Some of the statements included will apply regardless of age or activity although one would expect that most would have been addressed comprehensively by the time the pupil reaches Key stage 4. A number will clearly be more applicable to some activity areas than to others. Individual physical education departments need to review their own curriculum and the context within which they work and make decisions about when and where specific elements of the general requirements are to be taught or reinforced.

Promotion of physical activity and healthy lifestyles

Levels of physical activity

Exactly what is meant by 'teaching pupils to be physically active' is debatable. As with a great deal of the new Order, the wording leaves much to be desired. Pupils may be physically active in the short term, that is for the duration of the lesson or series of lessons, but this requirement is set in the context of healthy lifestyles, implying that what is expected is something rather more long term. Learning about the importance of being physically active, about the kinds of physical activity which will promote the sort of healthy lifestyle which is being encouraged, is not necessarily the same as ensuring that pupils are physically active in the short term.

Most research into the activity levels within physical education lessons suggests that, in many lessons, activity levels are unacceptably low. Where the activity takes place indoors, in relative warmth, as in the case of gymnastics, the temptation to talk too much at the expense of pupil activity is always present. This is not to say that purposeful talk is a waste of time. Some important elements of the knowledge and understanding needed to make informed decisions about subsequent activity interests can be covered effectively through the medium of the gymnastics lesson and this may be the most appropriate slot on the curriculum.

Posture and appropriate body use

Gymnastics is an excellent vehicle for the promotion of good posture and safe exercise habits.

Development of cardiovascular health, flexibility, muscular strength and endurance

Because gymnastics has the capacity to develop all of the above, to a greater extent than, for example, participation in some games, lessons should include the development of understanding about the factors which contribute to fitness and the kinds of activity which will develop them. For example, several pupils are performing bridges in response to a task. Two can be chosen to demonstrate how one is able to attain a higher position than the other, where the flexibility is, the extent to which the other could improve and how she might go about achieving this improvement.

One pupil can roll round a beam to finish in a held half lever position. The rest of the group find the ending impossible. This can be used to explain what strength is needed and in what muscles, and how strength is improved.

Personal hygiene

It is normal practice for teachers of Key stage 3 pupils to teach about personal hygiene. This often centres around the knotty issue of showers after lessons. It has to be said that the practice of showering seems, all too often, to be more related to matters of time, logistics and availability than to any real need! Pupils are well aware of this and, for many, school physical education experience is coloured by their intense dislike of rushed public showering, often when they have not been active enough to need a shower. This issue becomes higher profile where pupils' culture forbids or disapproves of public showering.

Development of positive attitudes

Conventions of fair play, honest competition and good sporting behaviours

Gymnastics, as will be discussed later, can contribute a great deal to personal and social education through use of particular teaching styles and specific activities. However, its contribution to the actual requirement quoted above will be less than that offered through the games programme which will be best placed to deliver this particular requirement.

Cope with success and limitations in performance

This, of course, implies that there will be success to cope with! One of the attractions of gymnastics is its capacity to offer success to a wide range of abilities and aptitudes. Skilful teaching can therefore enable all pupils to recognise and enjoy success while learning to understand why their performance may be limited in some areas.

Try hard to consolidate their performance

This implies that appropriate challenges are set. Pupils will be able to practise purposefully if the task is within their capabilities and offers some challenge. It is important that they can see some purpose in practising which means that they must know what the intended outcome of the practice is – mastery of a new skill; smooth performance of actions within a sequence; improved synchronisation with a partner, and so on.

Mindful of others and the environment

Gymnastics at Key stage 3 offers great opportunities for actively helping others over and above the need to respect and watch for others which is ever present.

Ensuring safe practice

Respond readily to instructions

This should have been established at Key stage 1 but will need revision at later key stages.

Recognise and follow relevant rules, laws, codes, etiquette and safety procedures for different activities or events

While this clearly has some relevance for gymnastics, its major emphasis is upon competitive activities. It will, however, be of importance if, for example, a gymnastics activity such as trampolining is introduced at Key stage 4, since there are quite specific safety procedures to be learned in addition to codes and rules if competitive events are involved.

Safety risks of wearing inappropriate clothing, footwear and jewellery, why particular clothing, footwear and protection are worn for different activities

Appropriate clothing is important both for safety in relation to the activity, for example not wearing skirts for gymnastics because of the danger of getting caught on the apparatus, and for safety in relation to keeping warm and preventing injury. It is the latter which is not always heeded sufficiently in schools. Beginning lessons in cold gymnasia dressed only in shorts and a T-shirt is not consistent with teaching that warm muscles function more effectively and safely. Wearing of tracksuits or at least sweaters should be encouraged at least for the early part of the lesson.

Lift, carry, place and use equipment safely

Gymnastics inevitably involves considerable handling of apparatus, some of which is heavy, awkward or both. There are therefore many opportunities for teaching safety in lifting and carrying. Safe use of apparatus should be a feature of every lesson and pupils should be increasingly capable of taking responsibility for their own safety, although the final responsibility, of course, remains with the teacher.

Warm up and recover from exercise

Pupils should understand general principles for warming up and recovering and also the specific requirements for an activity such as gymnastics. For example, the flexibility required for the performance of certain skill makes the preparation and warming of muscles particularly important.

What the general requirements do point to is the potential of physical education in general, and gymnastics in particular, to contribute to aspects of health-related work and to personal and social education. These two cross-curricular areas will be considered in this chapter.

Gymnastics and health-based physical education

There has been considerable debate about the advantages and disadvantages of teaching health-related issues as a discrete unit of work as opposed to allowing them to permeate all aspects of physical education. Unlike the Department of Education for Northern Ireland, which includes a specific compulsory health-related physical education unit at Key stage 4, the National Curriculum for England and Wales includes health-related matters as part of general requirements at all four key stages. Whether physical education departments choose to teach discrete units of health-related work or not, permeation of these issues throughout the activity programme is essential if learning is to be maximised and pupils are to understand how all the various physical education activities can contribute to a healthy lifestyle.

Gymnastics can contribute to this area in several ways:

★ as a tool for promoting understanding of the effects of different kinds of exercise on the body;

★ as an activity which incorporates health and fitness promoting actions and skills.

★ as a vehicle for the practical illustration of how the human body works.

Promoting understanding

The integration of health-based issues into the gymnastics lesson is key to long-term learning and understanding by pupils. The teacher must decide which aspects of this work are best addressed in the gymnastics context. For example, the warm-up phase of the lesson might be an excellent medium for ensuring that all pupils understand what happens to their heart rate when their activity level rises. It might not be the most appropriate medium for learning about the effect of sustained endurance activity on heart rate, since athletics activities could be a more effective teaching medium.

Gymnastics does involve cardio-vascular endurance, strength and flexibility, and it is often a revelation to pupils that activities such as gymnastics and dance are better developers of all three fitness components than are many games. As a result, there are many opportunities within gymnastics lessons for reinforcing understanding of the particular physical demands which specific actions make and of how easily the fitness to meet such demands can be acquired.

For example, some pupils will be able to use upper body strength to climb ropes, or for hanging and heaving activities using beams or bars. This creates a context in which the teacher can ensure that others understand why it is that one pupil finds this activity quite easy while for another it is impossible, and also what kinds of training would be needed to improve this particular fitness dimension.

Some pupils will be able to perform a bridge easily, or the splits. Understanding first that these require flexibility, second flexibility in specific muscles, e.g. the shoulders for a bridge, and third that warm muscles and daily stretching is needed to improve flexibility, can all be gleaned from a demonstration and quick explanation.

Health- and fitness-promoting activities

The incorporation of health- and fitness-promoting actions and skills into the lesson is not a problem within the gymnastics context. Giving sufficient time for such activities to have an effect may well be difficult, especially if lessons are short and units of work last half a term or less. Certainly a whole unit of work (A and B) at Key stage 3 would be needed to maximise the potential of gymnastics in this as in other areas. Nevertheless, much can be done to ensure that material is presented in a way which makes physical demands on the body.

Simply asking for repetitions of single actions or short action phrases increases the physical demands. For example, shoulder balance, stand up, half turn jump, return to shoulder balance, repeated three or five times, make considerable physical demands. Bunny hops or handstanding activities which are repeated and where pupils are asked to spend progressively longer on their hands, will make demands on muscular strength. Hanging between two ropes or beneath a bar in a half lever position demands upper body and abdominal strength. Repeating the position several times increases this demand.

Any repetition of short sequences, for example, balance, jump, turn, will develop both cardiovascular and muscular endurance.

Practical illustrations

Gymnastics can be a useful focus for practical illustrations of aspects of anatomy and physiology in the context of GCSE or A level work as well as for younger pupils, for example, knowing that the muscles which need to stretch for some warm-up actions are those down the back of the leg and that they are called hamstrings.

Gymnastics and personal and social education

The potential of physical education to contribute to personal and social education has long been recognised. HMI, writing in 1989, state that,

> *The major emphasis . . . should be on sharing and working together. Physical education produces many opportunities to contribute to pupils' personal and social and moral development, for example:*
>
> > *in the sharing and showing of work . . .*
> > *in co-operative activities . . .*
> > *in appreciating and accepting the different abilities and qualities of their peers . . .*
>
> *Pupils need to be encouraged to take some responsibility for their own learning and to be able to work with and help others. (DES/HMI 1989)*

Early National Curriculum documents produced by the now defunct National Curriculum Council also identified personal and social education as an important cross-curricular dimension.

> *Personal and social development involves aspects of teaching and learning which should perme-ate all of the curriculum. Whilst secondary schools may offer courses of personal and social edu-cation, it is the responsibility of all teachers and is equally important in all phases of education (NCC 1989).*

Gymnastics offers many opportunities for personal and social education, both through the content of the programme of study and through the potential provided by the varied teaching styles which may be employed. Much writing about personal and social learning makes reference to teaching strategies and, in particular, to the importance of creating a co-operative learning environment. It is therefore unfortunate that the latest National Curriculum Order uses the language of pupils 'being taught', when other language could have given clearer indications of a recognition of the potential of varied teaching and learning approaches. Nevertheless, as discussed in Chapter 5, it is clear that use by the teacher of a range of teaching approaches will not only help all pupils to benefit through recognising that learning styles differ, but will also maximise the potential of physical education to achieve a wide range of objectives.

Reciprocal teaching is often identified as the teaching style most clearly associated with fostering personal and social education because of its potential for involving pupils in peer tutoring (Williams 1993, Underwood and Williams 1991). For this to be successful, personal and social qualities required must be made explicit and discussed with pupils. This may seem to be a waste of valuable activity time, especially when physical education time is under pressure in some schools, however the long-term benefits of independent learning and working should far outweigh the short time needed to introduce this way of working.

Several approaches may be adopted. The following are two examples.

Example A

Explain to pupils that they will be working together and helping each other and discuss why this will help them to learn (i.e. more feedback from each other than would be possible for one teacher to provide for a whole group of pupils).

Ask them to discuss what will be important if they are to work together successfully. Ask them for their most important idea and list them on a flipchart/chalkboard.

A typical response might be:

★ be patient;

★ help each other;

★ be responsible;

★ be nice to each other;

★ praise each other;

★ don't be bossy;

★ don't shout at each other;

★ take turns.

Check that pupils understand the meaning of the terms used.

Display the negotiated list in the working area to serve as a reminder for subsequent lessons.

Example B

Explain to pupils that they will be working together and helping each other and discuss why this will help them to learn (i.e. more feedback from each other than would be possible for one teacher to provide for a whole group of pupils).

Split pupils into small groups and give each a set of cards with different personal qualities written on them (e.g. patient, helpful, selfish, rude, friendly, etc.).

Ask pupils to choose the three cards/qualities which they think will be most helpful if they are to work together.

List qualities and 'votes' for them on a flipchart/chalkboard.

Check that pupils understand the meaning of the terms used.

Display the negotiated list in the working area to serve as a reminder for subsequent lessons.

In addition to selecting a teaching style which fosters personal and social development, specific teaching strategies can also help or hinder. For example, if part of the agenda is to encourage respect for all members of the class, then the teacher's own actions are very important. Involving all pupils in the learning process can be fostered by asking questions of everyone, boys and girls, able and less able. It is helped by judicious selection of pupils to demonstrate. The teacher's comments on work produced or shown is also crucial, for example, in helping pupils to recognise each other's achievements (however small) rather than denigrating them.

The content and the context of gymnastics provide many opportunities for co-operative work. Because gymnastics is not inherently competitive, it offers a different kind of experience from that of games. Pupils need, at the most basic level, to share working space. They need to share equipment and apparatus. If learning is to be maximised it is important that particular groups do not colonise particular areas of the gymnasium or pieces of apparatus. By Key stage 3, pupils should be capable of managing choices, whether this is between a number of tasks or between apparatus of different heights/sizes or layouts. This will not be possible if, for example, a particular piece of apparatus becomes a 'no-go' area for the girls.

Work with a partner or as part of a larger group also demands effective co-operative working and recognition of each other's strengths and limitations.

CHAPTER 11

Workcards

Cards based on balance

Tasks for cards 1–6

★ How many of these balances can you perform?

★ Can you watch your partner and make sure that they look like the picture on the card?

★ Choose one balance that you can perform comfortably. Can you change the shape/position of your

Card 2

Card 3

Card 4

Card 5

Card 6

legs while performing the balance? Can you produce three variations? If not, choose another balance which makes this possible.

★ Can you make your balance the end of a forward roll? The end of a backward roll? Follow a handstand (or weight on hands action)?

★ Can you find a balance which can be both the start and the end of a forward or backward roll?

★ Can you use a spinning movement to link two of these balances? A twisting movement?

Tasks for cards 7–9

★ How many of these balances can you perform?

★ Can you find two different ways of getting into and out of each balance?

★ Can you perform these balances synchronising with a partner?

★ Can you find other balances of your own?

★ Can you perform these balances on any other piece of apparatus?

Card 7

Card 8

Card 9

Cards based on rolling

Tasks for cards 10–12

★   Can you copy these positions? Which are the easiest? Why are some difficult?

★   Look at these ways of ending a roll. How many can you do?

★   Can you help your partner with some of them?

★   Can you roll in any other direction to finish in these positions?

★   Which of these positions can you use as the beginning of a forward or a backward roll?

★   Can you follow the roll with a spinning action? A twisting action? A jump? Another turn?

Tasks for cards 13–14

★   Can you help your partner to perform this roll?

★   Can you perform this roll and then add another action to finish in an inverted balance?

★   Can you perform this roll and follow it with a twisting action? A spinning action? A jump? Another turn?

Card 10

Card 11

Card 12

Card 13

Card 14

Cards based on flight

Tasks for cards 15–16

★ How many of these jumps can you perform from the floor?

★ Which of these jumps can you perform with a single take off?

★ Which of these jumps can you perform with a double take off?

★ Which of these jumps can you perform from apparatus? (N.B. Landings from apparatus must be onto two feet.)

Cards based on stepping

Tasks for cards 17–18

★ Take your weight on your hands and come down to finish in one of the positions on the card.

★ Which of the other positions can you finish in?

Card 15

Card 16

Card 17

Card 18

★ How many positions can you go straight into from a handstand?

★ Can you find more than one way of coming down and finishing in your chosen position?

★ Can you come down from your handstand so that you can follow it with a spinning action? A jump? A twisting action? A turn?

Cards based on partner work

Tasks for cards 19–20

★ Can you perform the balances matching your partner as shown?

★ Can you perform the balances mirroring your partner as shown?

★ Can you link the three matching balances to make a matching sequence?

★ Choose one balance to begin your sequence and another to end it. Use jumps, rolls, twists, steps and spins to create a matching sequence to move from one balance to the other.

★ Which of these balances can you perform on apparatus? Can you adapt others?

Card 19

Card 20

Tasks for cards 21–24

★ Choose one of the balances on the sheets and perform it with a partner or help a pair to perform it.

★ If you wish, choose one of the balances and adapt it so that it can be performed by three people.

★ Find four other balances which you can perform as a pair or a three.

★ Choose one of the balances and using rolls, jumps, twists, stepping actions or spins, move into and out of the balance in a controlled way.

★ Having moved out of one balance, find a way of coming back together to perform a second balance.

Cards based on sequence development

Tasks for cards 25–27

★ Can you perform this sequence?

★ Can you help a partner to perform this sequence?

Card 21

Card 22

Card 23

Card 24

Card 25

Card 26

Card 27

Tasks for cards 28–32

★ Choose one of the single balance cards.

★ From this position can you,
   – roll sideways;
   – roll backwards;
   – roll forwards;
   – take your weight on your hands?

Tasks for cards 33–36

★ Choose one of the sheets with two balances on it. Make sure that you can perform both balances.

★ Can you link the two balances using a roll?

★ Can you link them using a jump?

★ Can you link them using a stepping action?

★ Can you link them using a spinning action?

★ Can you link them using a twisting action?

From this starting position can you:

1. roll forwards?
2. roll backwards?
3. handstand or cartwheel?

Card 28

From this starting position can you:

1. roll sideways to finish in the same position?
2. roll backwards onto one foot or to straddle?
3. lie flat and return to the same position slowly?

Card 29

From this starting position can you:

1. roll sideways?
2. roll backwards?
3. roll forwards?
4. take your weight on your hands?

Card 30

From this starting position can you:

1. roll backwards?
2. get onto feet and stretch jump?
3. get onto knees, crouch and headstand?
4. roll backwards onto one knee?

Card 31

From this starting position can you:

1.   roll forwards?
2.   come down to a straddle stand?
3.   come down and roll backwards?
4.   come down and stretch jump?

Card 32

Can you hold these two balances?

Now join the two balances together

1.   using a roll
2.   using a spinning movement
3.   taking your weight on your hands

Card 33

Now hold these two balances.

Now join these two balances together

1. using a backward roll
2. using a turning jump
3. using a forward roll and a jump

Card 34

Can you hold both these balances?

Now join the two balances together

1. using a roll
2. using a spinning movement
3. taking your weight on your hands

Card 35

Now hold these two balances.

Now join these two balances together

1. using a forward roll
2. using a handstand
3. using a backward roll and a jump

Card 36

Tasks for cards 37–38

★ Choose one of the sheets with five balances on it. Choose four of the balances and make sure that you can perform them.

★ Use your four balances to make the framework of a sequence.

★ Use different actions to join the balances together. You could use spins, jumps, twists, steps, cartwheels, and so on.

★ Choose two of the balances and perform them on apparatus. Plan and perform a sequence which involves two apparatus balances and two floor balances.

Can you hold these five balances?

Choose four of them to join together to make a sequence.
Use different actions to join the balances together.
You could use spins, rolls, steps, cartwheels, handstands, jumps.

Card 37

Can you hold these five balances?

Choose four of them to make a sequence.
Use different actions to join the balances together.
You could use spins, rolls, steps, cartwheels, handstands, jumps.

Card 38

Cards based on specific skills

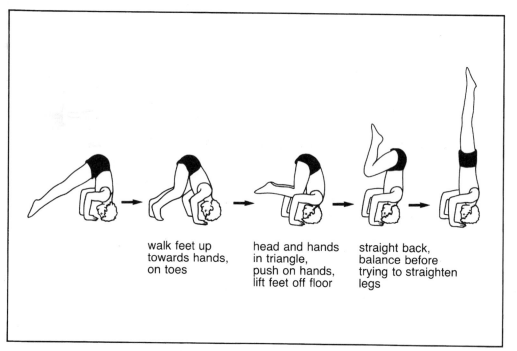

walk feet up
towards hands,
on toes

head and hands
in triangle,
push on hands,
lift feet off floor

straight back,
balance before
trying to straighten
legs

Card 39

reach forward with
hands, and place
on floor

hands shoulder width
apart, arms straight,
fingers forward
swing first leg up

look at floor
keep back straight

Card 40

chest towards knee,   legs wide apart
strong lift with first leg

Card 41

shoulders ahead of hands
to begin roll

kick up into   into good handstand   tuck head well in,   reach forward with arms,
handstand   before trying to roll   round back   stand up once weight is
                                                                 on balls of feet

Card 42

Bibliography and further reading

BAALPE (1995) *Safe Practice in Physical Education*. BAALPE

Bilbrough, A. and Jones, P. (1973) *Developing Patterns in Physical Education*. London: Hodder and Stoughton

Board of Education (1909) *The Syllabus of Physical Exercises for Schools*. London: HMSO

Brown, G. and Wragg, E.C. (1993) *Questioning*. London: Routledge

Burrows, W. (1991) *Progressions in Gymnastics (Video and Notes)*. Brighton: Brighton Polytechnic

Carroll, B. (1994) *Assessment in Physical Education*. Brighton: Falmer Press

Cope, J. (1967) *Discovery Methods in Physical Education*. Walton-on-Thames: Nelson

DES/HMI (1989) *PE 5–16 Curriculum Matters*. London: HMSO

DES (1990) *National Curriculum Physical Education Working Group: Interim Report*. London: DES

DES (1992) *Physical Education in the National Curriculum*. London: HMSO

DNH (1995) *Sport: Raising the Game*. London: DNH

Groves, R. (1973) 'Sporting doubts about educational gymnastics', in *British Journal of Physical Education*, **14**, 4

Harlen, W. (1991) National Curriculum Assessment: increasing the benefit by reducing the burden, in 'Education and change in the 1990s', *Journal of the Educational Research Network of Northern Ireland*, **5**, February 3–19.

London County Council (1965) *Educational Gymnastics: A Guide for Teachers*. London: LCC

Mace, R. and Benn, B. (1982) *Gymnastics Skills*. London: Batsford Books

Maulden, E. and Layson, J. (1979) *Teaching Gymnastics* (2nd edn). London: MacDonald and Evans

Ministry of Education (1952) *Moving and Growing*. London: HMSO

Ministry of Education (1953) *Planning the Programme*. London: HMSO

Morison, R. (1956) *Educational Gymnastics*. London: PEA

Morison, R. (1969) *A Movement Approach to Gymnastics*. London: Dent

Mosston, M. and Ashworth, S. (1986) *Teaching Physical Education*. Toronto: Merrill

Munrow, D. (1963) *Pure and Applied Gymnastics*. London: Edward Arnold

National Curriculum Council (1989) *Circular No 6 The National Curriculum and Whole Curriculum Planning; Preliminary Guidance*. York: NCC

Sabin, V. (1993) *Teaching Gymnastics Skills* (Volumes 1, 2 and 3). Val Sabin Publications

Smith, B. (1991) 'The Contribution of Gymnastics to Health-based Work', in *British Journal of Physical Education*, **22**, 3, 23–25

Smith, T. (1984) *Gymnastics – A Mechanical Understanding*. London: Hodder and Stoughton

Spackman, L. (1995) 'Assessment in Physical Education', *British Journal of Physical Education*, **26**, 3, 32–34

Underwood, M. (1990) *Agile*. Thomas Nelson

Underwood, M. and Williams, A. (1991) 'Personal and Social Education through Gymnastics', in *British Journal of Physical Education*, **22**, 3, 15–19

Williams, A. (1993) 'The contribution of physical education to personal and social development', in *Pastoral Care in Education*, **11**, 1, 21–25

Williams, J. (1973) *Themes for Educational Gymnastics*. Wakefield: Lepus Books

Wright, J. (1980) *Association of Principals of Women's Colleges and Physical Education Conference Paper*, Nonington College

Index